FALLS FROM HEIGHT

A Guide to Rescue Planning

LOUI McCURLEY

WILEY

Published by John Wiley & Sons, Inc., Hoboken, New Jersey
Published simultaneously in Canada

For general information on our other products and services or for technical support, please contact our Customer Care Department within the United States at (800) 762-2974, outside the United States at (317) 572-3993 or fax (317) 572-4002.

Wiley also publishes its books in a variety of electronic formats. Some content that appears in print may not be available in electronic formats. For more information about Wiley products, visit our web site at www.wiley.com.

Library of Congress Cataloging-in-Publication Data:

McCurley, Loui, 1965–
Falls from height : a guide to rescue planning / Loui McCurley.
pages cm
Includes index.
ISBN 978-1-118-09480-8 (hardback)
1. Industrial safety. 2. Falls (Accidents)–Prevention. 3. Rescue work. I. Title.
T55.3.H45M35 2013
658.3'82–dc23

2012049880

Printed in the United States of America

10 9 8 7 6 5 4 3 2 1

Psalm 86:13

. . . for your love for me is very great.
You have rescued me from the depths of death.

CONTENTS

FOREWORD

An accident happens. Someone's life is in danger and for most of us the immediate reaction is to dial 911 or another local rescue service call-out number. That is almost always the right thing to do but for many workplace accidents, whoever responds may not be prepared to solve the problem quickly and safely, if at all.

As a 40-year veteran of the rescue service, I know well that all responders want to help and will often risk their own lives to save the lives of people they do not even know. Even if they are prepared to help, the time it may take for responders to reach the accident site, much less access the victim, can be too long in many locations and environments.

Emergency services dispatchers can send fire departments, ambulances, and law enforcement most anywhere in the United States. Areas that have environments that are difficult to access often have plans for calling out specialty rescue teams that are trained and equipped for mountain, cliff, avalanche, cave, mine, tunnel, wilderness environment, river, surf, and a host of other specialty technical rescues. In addition there are military responders like the Coast Guard and Air Force pararescue jumpers, ready to respond. Even so, response to difficult locations is often too late for the victim.

Most of us might think that an industrial work site would not be one of the aforementioned situations. We might assume that a workplace can rely on local public emergency responder services, particularly in the case of a "simple" fall off a work surface that is caught by fall protection equipment. The truth is that the employer has a duty to provide rescue in a timely manner whenever an employee is injured or in a situation of threat to life. Not only are employers required to provide proper safety protection, they must also have an effective rescue plan. Just calling 911 is not an effective rescue plan. It may not be the fall that gets you, or the almost sudden stop as one's fall arrestor deploys, but simply the wait for someone to get you down.

Falls from Height: A Guide to Rescue Planning is about responding to a fall safely and quickly, be it by your own skills and training, a co-worker, an in-plant emergency team, or an offsite rescue service. This book covers the regulatory rules, planning, and training needed to prepare the employer to get the rescue job done.

Many have written about protecting workers from falls but Loui McCurley takes fall protection to the next step: How do you get the person hanging in a harness from a lanyard 50 feet in the air over a rack of steam pipes, the person caught in a net 150 feet in the air and below a stadium roof, the person unconscious 300 feet up a tower on a caged ladder climb, or the person hanging from a cell phone monopole tower's climbing pegs down quickly? More importantly, how do you do so without further injury, and with safe techniques for the rescuers?

Loui McCurley has been involved in technical rope rescue since 1985. It was her interest in mountaineering as a sport that has since led to a lifelong career interest in rescue from height. I first worked with Loui when she was testing (read: "breaking") rescue equipment in simulated field situations in order to find the best pieces of gear for the job. From the outset, Loui had a passion for finding answers and sharing what she learned with other rescuers. Since the early 1990s she has been the moving force behind finding great presenters for the International Technical Rescue Symposium, which is dedicated to sharing information among rescuers.

Also in the 1990s, Loui and others began to discuss the formation of an organization to help promote safety among suspended workers at height. While the Industrial Rope Access Trade Association existed already, this was a trade association for Rope Access *companies*. A worker-oriented organization had yet to be founded. Loui quickly took off with that project and organized a small group of like-minded individuals to found the Society of Professional Ropes Access Technicians (SPRAT). SPRAT represents the interests of the individuals who work at height, and has certified thousands of rope access workers who are now performing work at height all over the world.

Through her commitment to safety, supported by her paying job as Senior Vice President of Pigeon Mountain Industries, Loui has become involved in educating regulators, equipment manufacturers, and employers of the safety advantages of utilizing rope access for work at height, and the value of certification as a means of skills verification. This path has led Loui through a tangled web of national and international fall protection regulations and standards, and directly back to the place where she started: a place of deep concern and passion for safety in work at height.

In this one-of-a-kind text, Loui shares her wealth of experience to provide the reader with a thorough treatise on the subject of preparing for rescue after a fall in the workplace. In it, she covers all the bases, from who is responsible for what, to hazard awareness and mitigation, review of existing response protocols, new preplans, and specific recommendations on how to get it done. This is a book you can really use.

Stephen E. Hudson, President
Pigeon Mountain Industries

PREFACE

Whenever fall protection methods are used, the employer has a moral, ethical, and sometimes legal obligation to provide for prompt rescue (including self-rescue) in case of a fall.

To effectively prepare for prompt rescue, an employer must evaluate and understand his or her specific rescue needs and make appropriate preparations in advance. This book is designed to assist safety professionals whose job it is to plan for the safety of at-height workers, and for those whose responsibility it is to prepare for and perform post-fall rescue. This is a multi-faceted book, and may be used in any number of ways, depending on the needs of the reader. Some will find it advantageous to read the text cover-to-cover. Others may find it more beneficial to customize the approach to the text, and perhaps even read it in a different order than it is presented.

HOW TO USE THIS BOOK

This book is designed for simplicity of use, maximum versatility, and to facilitate understanding by a variety of users ranging from the most skilled of field practitioners to the corporate executive of health and safety. The text is organized into chapters, each of which addresses a specific aspect of rescue in the workplace. While each chapter stands alone on a given topic, concepts in some chapters do build upon concepts that may have been initially introduced in another.

The practitioner who might wish to gain a thorough understanding of all aspects of the rescue equation may benefit from reading the book in its entirety, from front to back, while the industry professional who wishes only to brush up on a few things might prefer to pick and choose chapters most applicable to his need.

For ease of reference, the 11 chapters are grouped into parts containing related topics. The parts and chapters are as follows:

Part I: Regulatory Considerations, Rescue Plans, and Developing a Rescue Capability

This part is a must-read for managers and supervisors. It validates the need for a rescue program and provides guidance toward achieving maximum results with minimal resources. It contains three chapters:

- Chapter 1: The Reason for Rescue. This chapter provides an overview of fall hazards and reviews the types of falls that occur in industry. It also offers guidance toward employer responsibilities, explains the role of voluntary consensus standards in relation to regulatory requirements, and offers insights on the comprehensive managed fall protection plan.

- Chapter 2: Developing a Post-Fall Rescue Plan. This chapter provides guidance on how to approach the development of a rescue plan, including how to determine whether a rescue plan is required, what the rescue plan should contain, and consideration of other hazards. This chapter also offers guidance on evaluating a response team and coordination with external resources.
- Chapter 3: Program Development and Incident Management. In this chapter, the reader will begin to explore the mechanics of the rescue process and gain a clearer understanding of the differences between personal escape, co-worker assisted response, and professional rescue. Roles and responsibilities are defined to establish a framework for selecting personnel, training, and managing an incident.

Part II: Equipment, Systems, and Skills

This part is for the field practitioner who really wants to gain the skills and knowledge to perform rescue at height. It explores requirements for fall protection equipment as compared with rescue equipment, and describes basic skills that every rescuer should be able to perform.

- Chapter 4: Equipment for Co-worker Responders. This chapter explores selection, care, and maintenance of equipment that is worn or used by the rescuer for safety and rescue.
- Chapter 5: Rescue Equipment and Systems. In this chapter, the reader gains a deeper understanding of equipment that is appropriate for life safety and rescue applications, specifically that which is used for rigging. Attention is given to pre-engineered systems as well as components.
- Chapter 6: Personal Skills for Rescuers. Here the reader is introduced to the basic skills that establish a foundation for rescue. Methods for anchoring, descending, ascending, and patient care are addressed.

Part III: Putting It All Together

This part is of interest to both safety managers and field practitioners. It introduces systems and techniques for rescue, offers guidance toward bringing the big picture into focus, and guides thought processes toward specific hazards and industry considerations.

- Chapter 7: Principles of Rigging. This chapter offers a cursory understanding of the physics that affect rescue systems and how to use those concepts in harmony with the equipment introduced in Part II to achieve rescue-related tasks such as anchoring, lowering, raising, and re-directing a load.
- Chapter 8: Single Rescuer Techniques. Building on the concepts in Chapter 7, this chapter introduces specific methods and techniques that can be used by

one co-worker to rescue another in almost any environment. Step-by-step guidance is provided to enhance training.

- Chapter 9: Group Rescue Methods. This chapter also expands the concepts found in Chapter 7, but this time with specific methods and techniques that are more conducive to a multi-person response. Again, step-by-step guidance is provided to enhance training.

Part IV: Successful Workplace Planning

This part is primarily of interest to safety managers, although experienced field practitioners will benefit from understanding and applying the information in this part as well. Some particular nuggets that are applicable to various industries may be found in Chapter 10.

- Chapter 10: Industry-Specific Considerations. While rescue from fall protection shares some similarities (such as gravity) regardless of environment, each unique job involves its own distinctive challenges and hazards based upon the type of work being performed, the environment, and other factors. This chapter provides insight into assessing hazards in several common workplace environments.
- Chapter 11: Practical Workplace Application. This chapter is a step-by-step guide for implementing the contents of this book. It contains forms, checklists, and guidelines that both the manager and the practitioner can use to establish, perform, and maintain post-fall rescue capabilities in a workplace.

Again, it is important to note that while the chapters in this book attempt to offer a thorough treatise on their respective subjects, the information from beginning to end is intertwined and interrelated.

YOUR SUCCESS

It is my deep desire that this book aids you along the way toward greater safety in your workplace environment, whether you are a worker, a manager, a supervisor, an employee, or an employer. If you have suggestions or recommendations for future revisions of this book, or of companion works, I welcome you to share those with me as I seek to continue in my mission to equip workers at height with tools for safety.

Stay safe!

Loui McCurley

ACKNOWLEDGMENTS

I would like to thank those who have, knowingly or unknowingly, contributed to the development of this text. First, my supportive husband Bob and sons William and Ian for their encouragement and respective votes of confidence, and to my mom for instilling a love of books. A special thanks to Bob, especially, for not being afraid to tell me when parts of the text were gobbledygook, and for assisting with moral support, encouragement, and editing.

I would also like to thank my talented team of content critics who so valiantly waded through reams of material to weed out extraneous information, correct errors, and recommend improvements: John Peleaux, for sharing decades of rigging and technical expertise; Barbara Bielaski, for her wealth of regulatory and industry knowledge related to fall protection; and Mike Roop, for his rescue management and industrial/confined space experience. Others, too, without whose specific knowledge and contribution the content would be lacking: Michael Stewart, who contributed safety management and engineering experience in the field of wind energy; Michael Seal, for lending safety management experience in bridge inspection and engineering; John Paul Jones, whose tower safety and rescue expertise was significant; Amos Yost, for sharing his unique expertise related to agricultural facilities and grain bin rescue; and Steve Hudson, for years of mentoring prior to this work, as well as for helping me to put some finishing touches on the text.

Finally, I would like to thank Robert McCurley and William Clem for the many hours of photography assistance they provided for this book, as well as those who facilitated the provision of photos, including Pigeon Mountain Industries; Vertical Rescue Solutions; Garin Wallace and Donald Enos of Seattle Manufacturing Co.; Keith Jones of Heightec, Ltd.; Dan Henn of Reliance Industries, LLC; Doug Knap of Tractel, Inc.; Michael Seal of Burgess & Niple; J.P. Jones of Safety LMS; Climb-Tech; Beal SA; Dawn Wilson; and John Peleaux of Innovative Access, Inc.

L.M.

PART I

REGULATORY CONSIDERATIONS, RESCUE PLANS, AND DEVELOPING A RESCUE CAPABILITY

CHAPTER 1

THE REASON FOR RESCUE

By the end of this chapter, you should understand:

- The role of regulatory requirements in fall protection.
- The role of voluntary consensus standards in relation to regulatory requirements.
- How to approach forming a comprehensive managed fall protection plan.
- The hierarchy of fall protection.
- Common types of fall protection.

Early one December morning, on a swing stage scaffold suspended 50 feet above an ore dock, two workers performed maintenance work. In preparing to move the scaffold from one location to another, they discovered that one of the cables was kinked. As one of the workers worked to remove the kink, a strap that secured the swing stage to the dock let loose, causing one end to swing out from the dock. The worker was launched from the scaffold, striking the base of the dock 50 feet below before falling into the water. His surprised co-worker reacted as quickly as possible, extracted himself from the scaffold, and rushed to the base of the dock, searching for some way to help his companion. Spotting an extension cord nearby, he used it to lasso his companion's arm as he called for help. He did his best to hold his co-worker's head above water with the makeshift rescue line until rescuers were finally able to retrieve the employee from the water, using a boat. But it was too late. The fallen employee was not breathing at the time of rescue, and was later pronounced dead.[1]

Organizations that employ workers at height have an obligation to protect those workers who are at risk from falling during the course of their work. This is true whether the employee is working in a manufacturing facility, construction, transportation, agriculture, or any other industry. Being protected from a catastrophic fall, however, is just the beginning. Post-fall rescue is imperative whether the worker has impacted the ground or a lower level, or remains suspended in his harness.

[1]United States Dept of Labor; Occupational Health and Safety Administration, http://www.osha.gov/pls/imis/accidentsearch.html, Accident: 200514313: Employee Is Killed In Fall from Scaffold into Water; Report ID: 0552700. Event Date: 12/06/2006.

Falls from Height: A Guide to Rescue Planning, First Edition. Loui McCurley.

Figure 1.1 What does it feel like to be suspended after a fall?

Anyone who might harbor even the slightest doubt as to the importance of post-fall rescue for a suspended worker should engage in this simple little exercise:

1. Don a harness of the type that you/your workers normally wear.
2. Connect a lanyard to the harness in the manner in which it is normally used.
3. Connect the lanyard to an anchor that will allow you be suspended with your feet just at floor level.
4. Lift your feet so that you are hanging in the harness, as shown in Figure 1.1.

Within about 60 seconds, the importance of post-fall rescue will become crystal clear, without you even having suffered the effects of a fall!

Caution! This exercise should be attempted only under proper supervision by a qualified individual who is able to ensure immediate rescue. Always obtain your doctor's permission before engaging in this type of exercise.

REGULATORY REQUIREMENTS

In the United States, the obligation to protect employees stems from the Occupational Safety and Health (OSH) Act of 1970, which acceded regulatory authority for the workplace to the Occupational Safety and Health Administration (OSHA) and assigned them with a number of specific regulatory functions, among them:

1. Setting standards, and
2. Conducting inspections.

OSHA's specified purpose is "to assure . . . every working man and woman in the Nation safe and healthful working conditions and to preserve our human resources."[2]

As part of its mission, OSHA conducts inspections and issues citations to employers who do not provide adequate safety for their employees. Different types of violations can result in varying levels of financial penalties, with greater fines resulting from more serious violations. Violations deemed to be "willful" or "repeat" in nature can carry maximum fines up to ten times higher than serious violations. A *willful violation* is one committed intentionally, with foreknowledge or voluntary disregard for the law's requirements, or simply with indifference to worker safety and health. A citation for a *repeat violation* is issued when there have been previous citations for substantially similar violations at any work site owned by the same company.

The OSH Act was passed December 29, 1970, but OSHA did not come into existence until 1971. Since its inception, OSHA has developed numerous federal standards relevant to workplace safety and health. The standards cover four specific industries—construction, maritime, agriculture, and general industry—and are enforced throughout the United States. The term "standard" means something completely different when referring to OSHA standards as compared with voluntary consensus standards such as those developed by accredited standards-developing organizations. Because OSHA has enforcement authority, the standards (sometimes called rules or regulations) it sets have the force of law.

Employers who function under the laws of the United States are required to provide a safe and healthful workplace, as defined by these regulations, for their employees. OSHA monitors compliance to these regulations through inspections, whistle-blower programs, reporting requirements, and post-incident follow-ups.

Even in areas where OSHA has not identified a specific standard to address a particular hazard, employers are held responsible for employee safety under the Occupational Safety and Health Act's *general duty clause*. The general duty clause states that each employer "shall furnish . . . a place of employment which is free from recognized hazards that are causing or are likely to cause death or serious physical harm to his employees."[3] Falls are a recognized hazard, ranking among the top 10 federal standards with the most repeat violations cited by OSHA.

[2]OSH Act, Public Law 91-596, 84 STAT. 1590. 91st Congress, S.2193, December 29, 1970, as amended through January 1, 2004.

[3]OSH Act Section 5(a)(1).

Most other countries have some agency that serves a function similar to that of OSHA. In the United Kingdom it is the Health and Safety Executive (HSE); in Australia it is the National Occupational Health and Safety Commission (NOHSC); and in other countries there are other, similar agencies with other, similar acronyms. The regulations in each country differ from one another, so it is incumbent upon an employer working in a foreign country to become familiar with, and adhere to, the regulations of the country in which he is working.

Regulatory discussion is not a major focus of this book, but we will touch upon it to some degree as a foundation for the systems and methods discussed herein. Because this text is written in the United States, when the concept of governmental regulation of health and safety is referred to, it will be in context of OSHA. This is not to suggest that all readers from all countries should follow OSHA, it is only to serve as a point of reference. Readers from other countries are encouraged to further research their own regulatory authority (or authorities), with consideration to the concepts presented here.

Keeping workers safe is not just the law—it is also a good idea from a practical, economical standpoint. In 2009 alone, the Bureau of Labor Statistics logged more than 200,000 workplace falls. This, in turn, translates to billions of dollars in lost time, workers' compensation, litigation, medical costs, and so on. There is also the matter of worker morale to consider. No one wants to head off to work each day not knowing whether he or she will live to see another. Good employers know that safe, healthy workers are motivated, productive workers.

The category of potential workplace emergencies addressed by this text is that of fall-related incidents. Although we will briefly touch on some other specific hazards, employers with high risk exposure in these or other areas will want to consult industry-specific resources for solutions relevant to their particular application.

The majority of workplace falls are relatively simple slips and trips such as might occur on any walking surface. These types of falls are typically not catastrophic but are notable in that they may result in injury and/or time lost from work. These types of falls are important, but they are not the focus of this book.

Beyond the hazards of slips and trips on the horizontal dimension, many employers also employ individuals who work in or near the vertical dimension. This may include suspended work, scaffolds, elevated construction, large machinery, catwalks, container tanks, and other raised surfaces. Falls in these environments are much more likely to result in significant injury or even death. Falls that occur in this context, and more specifically, rescue of persons who fall while working in these types of environments, will be the focus of this book.

PROTECTING EMPLOYEES AT HEIGHT

OSHA regulations recognize that different employees may possess different levels of responsibility in the workplace. Workers may be held to different expectations, depending upon what level of responsibility the employer has bestowed upon them.

Throughout the regulations, the designations of Authorized, Competent, and Qualified are used to describe the differences. Although not all users of this manual will adhere to OSHA standards, these designations will be used throughout, as they are particularly useful in discussing different levels of responsibility.

As a general rule, the term "Authorized Person" refers to an individual who is approved or assigned by the employer to perform a specific type of duty or duties or to be at a specific location or locations at the jobsite. The term "Competent Person" is used in reference to one who is capable of identifying existing and predictable hazards in the surroundings or working conditions that are unsanitary, hazardous, or dangerous to employees, and who has authorization to take prompt corrective measures to eliminate them.

The term "Qualified Person" is a bit more of a loaded term. In regulatory-speak, it means "one who, by possession of a recognized degree, certificate, or professional standing, and/or who by extensive knowledge, training, and experience, has successfully demonstrated the ability to solve or resolve problems relating to the subject matter, the work, or the project." At this writing, there is a slight but meaningful distinction between the construction rules and the general industry rules in the United States. The powered platform (general industry) section states "one who, by possession of a recognized degree, certificate, or professional standing, *and* who by extensive knowledge . . ." while the construction rule states "one who, by possession of a recognized degree, certificate, or professional standing, *or* who by extensive knowledge . . ." (emphasis added by author).

Some OSHA standards present additional specific requirements that must be met by the Authorized, Competent, or Qualified Person(s), and the same or similar terms are also used by many standards throughout the world, usually in a similar manner. These concepts are foundational when planning any work at height, and any written documentation should specify what level of authority/responsibility the employer has designated to specific individuals for specific functions.

It is entirely possible—and even quite likely—for a person to be designated by their employer as an Authorized, Competent, or Qualified Person for one particular aspect of a job (such as fall protection, hazardous materials, or machinery operation), but not for other areas that pose different hazards or that require compliance to different regulations. On the other hand, it is possible that a given person might possess all of the experience and training necessary to allow him or her to be designated by the employer as an Authorized, Competent, or Qualified Person for every aspect of the entire jobsite. The more likely scenario is somewhere in between. A given person may be designated in one or more areas of the job as a Competent Person and as an Authorized Person in other aspects of the job.

In the United States, there is no single OSHA standard that establishes all of the specific responsibilities of Authorized, Competent, and Qualified Persons, but references to these appear throughout OSHA rules, standards, directives, and interpretations. Some of the OSHA standards list very specific requirements for qualification, tasks, and/or responsibilities, while others make only passing references. It is imperative that employers and safety professionals diligently research the regulation(s) that might apply to their type of work and environment to ensure that all requirements have been met for all levels of their employees.

Understanding employer and governmental regulation as it applies to fall protection for employees can be difficult, and this difficulty is compounded by the many different types of work and different kinds of fall hazards that may be present in the workplace. Additionally, OSHA has what it calls *duty requirements* that specify when an employer must have fall protection and other requirements that describe the criteria and practices that must be followed for each type of fall protection system used. However, it is simply not possible to organize fall protection, much less post-fall rescue, into a neat set of specific rules that fit all situations. OSHA acknowledges this difficulty when it separates its rules for fall-protection into one set of requirements for general industry, and another set of requirements for construction.

Federal rules for fall protection in the United States' construction industry are addressed primarily by OSHA in 29 CFR 1926.500 to .503, while fall-protection rules for general industry are found primarily in 29 CFR 1910.23 and 29 CFR 1910.132. It is important to note that it is the work being performed, not the job title, that determines which rules apply. Therefore, construction trade workers who are performing maintenance or general industry work would be held accountable to the applicable parts of the general industry standards found in 29 CFR 1910, and vice versa. In addition, 29 CFR 1910.66 has rules for fall protection when working on powered platforms, including both the duty to use fall protection and the criteria for personal fall arrest systems.

In accordance with U.S. federal regulations, as a general rule fall protection must be provided if an employee can free fall 4 feet or more onto a lower level in general industry or 6 feet or more onto a lower level in construction. This includes (but is not limited to) the following situations:

- When an employee is on a walking/working surface that has an unprotected side or edge.
- When an employee is constructing a leading edge.
- When an employee may fall through a hole in the walking/working surface.
- When an employee is working on the face of formwork or reinforcing steel.
- When employees are on ramps, runways, and other walkways.
- When employees are working at the edge of an excavation, well, pit, or shaft.
- When employees are working above dangerous equipment (even employees working fewer than 4–6 feet over dangerous equipment must be protected).
- When an employee is performing overhand bricklaying and related work.
- When an employee is performing roofing work on low-sloped roofs.
- When an employee is working on steep roofs.
- When an employee is engaging in precast concrete erection (with certain exceptions).
- When an employee is engaged in residential construction (with certain exceptions).

Already you can see that the employer has a big job on his or her hands when it comes to figuring out which regulations apply to his or her workers, and when. The complexity is further compounded as fall protection within these categories is then divided into several different subparts, depending on the nature of work being undertaken. There are separate guidelines, for example, for fall protection during work on scaffolds, on certain cranes and derricks, in tunnels, on stairways and ladders, during steel erection, and so on. Some of these have slightly different height limitations, and there are even complete exclusions in some instances.

The logic behind these variations can seem elusive, until you understand the scope, breadth, and complexity of the OSHA mission. OSHA adopted many of its original fall protection standards (like the current subpart D) from national consensus standards. Let's use the "4 foot rule" as an example. OSHA's fall protection rule in general industry requires fall protection for employees whenever they are working at heights in excess of four feet. This rule, which differs from other corresponding rules in OSHA regulations, originally came from a 1932 American National Standards Institute (ANSI) standard, A12.1. ANSI A12.1 has since been updated (A1264), and much work has been done to substantiate the survivability of falls greater than 4 feet—yet OSHA's 4 foot rule remains.

This story illustrates the fact that although national consensus standards are reviewed periodically, and many have been updated, corresponding OSHA regulations have not necessarily followed suit at the same speed. It is much more difficult to change an OSHA rule than to change an industry standard. Revised or new OSHA standards are subject to a stringent rulemaking process (including public participation) which, along with active lobbying from special interest groups, can make updates a lengthy and tedious process. As a result, sometimes regulations are not updated at the same rate of progress as technology.

As this book goes to press, OSHA is revising its general industry fall protection rules to consolidate the rules in subparts D and I. It is hoped that this will simplify matters and help facilitate compliance.

STATE PLANS

Employees who work for state and local governments in the United States are not covered by federal OSHA, but the Occupational Safety and Health Act does mandate protection for these workers in states that have an OSHA-approved state program, usually referred to as a *state plan*. A state plan is a job safety and health program that is similar to OSHA regulation, but operated on a state rather than federal level. OSHA approves and monitors state plans to ensure that they provide adequately for employee health and safety. State plans, by definition, may implement rules that differ from federal OSHA requirements, provided that those rules are (to use OSHA's words) "at least as effective" as the corresponding federal rules.

State plans may address one or more areas of safety that correspond with federal OSHA regulations, so it is important that the employer take all potentially applicable rules into consideration, and strive for compliance with whichever

TABLE 1.1 U.S. States and Related Regions with Approved State Plans

Alaska	Maryland	Puerto Rico
Arizona	Michigan	South Carolina
California	Minnesota	Tennessee
Connecticut[a]	Nevada	Utah
Hawaii	New Jersey[a]	Vermont
Illinois[a]	New Mexico	Virgin Islands[a]
Indiana	New York[a]	Virginia
Iowa	North Carolina	Washington
Kentucky	Oregon	Wyoming

[a]Plans cover public sector (state and local government) employment only.

regulation(s) are most conservative—that is, those that are most protective of the worker. More information can be found at www.osha.gov.

At this writing, the regions listed in Table 1.1 have state plans. Four these states (Connecticut, Illinois, New Jersey, New York) and one U.S. territory (Virgin Islands) have OSHA-approved plans that cover public sector (state and local government) employees only. Private sector workers in these four states and the Virgin Islands are covered by federal OSHA.

Where state plans exist, typically they adopt standards identical to the corresponding federal ones. Those that do vary will often also promulgate additional standards covering hazards not addressed by federal standards. State plan states must conduct inspections to enforce their standards, include public (state and local government) employees, and operate occupational safety and health training and education programs. In addition, most of these states provide free on-site consultation to help employers identify and correct workplace hazards.

OSHA requires federal employees to be protected by an effective and comprehensive safety and health program. Executive Order 12196, Section 1–201, and 29 CFR 1960.16 require federal agencies to adopt policies and procedures necessary to provide a level of protection equivalent to that provided by OSHA standards and regulations. Although OSHA does not fine federal agencies, it does monitor federal agencies and responds to workers' complaints. The United States Postal Service (USPS) operates under OSHA regulations.

At this writing, categories of workers not covered by the OSH Act include: those who are self-employed; immediate family members of farm employers who do not employ outside employees; and those who are regulated by another federal agency (for example, the Mine Safety and Health Administration, the Federal Aviation Administration, or the U.S. Coast Guard).

It is the responsibility of the employer to be aware which rules and regulations apply to a given workplace. Where multiple potentially applicable rules appear to be contradictory, it is always best to take the most conservative approach. The scope of this text will discuss only federal OSHA rules for general industry and construction in a very broad sense. State-specific variances from OSHA will not be addressed, nor will variations between countries. Do your homework and be very sure that your workplace health and safety program complies with the requirements of the country and jurisdiction in which you are working.

STANDARDS

Although OSHA has the statutory authority to develop, promulgate, and enforce occupational safety and health standards, they do not do so in a vacuum. The federal government recognizes the value of the consensus process in safety and health standards. In fact, when OSHA first started, it was required (under Section 6a of the act) to adopt national consensus standards for the first 2 years of its existence without rulemaking. This contributed greatly to the body of knowledge that was adopted at that time.

Today, OSHA still tends to nurture cooperative working arrangements with private organizations that develop voluntary national consensus standards on topics related to occupational safety and health. In fact, the National Technology Transfer and Advancement Act (NTTAA)[4] directs agencies to use national consensus standards instead of writing their own when possible. OSHA takes this even a step further: if the agency deviates from national consensus standards it is required under Section 6(b)(8) of the act to explain why.

OSHA's recognition of and cooperation with national consensus standards can be helpful to the employer because OSHA fall protection regulations tend to be very performance-oriented. That is, OSHA defines relatively broad, sweeping objectives for protecting employees from fall hazards. By necessity, methods for applying and achieving those objectives will vary depending on the environment and work-type, among other things, and OSHA appropriately leaves this topic to more narrowly focused industry-specific consensus standards such as those developed under the ANSI consensus procedures.

ANSI is an accrediting organization that oversees the development of standards by member organizations such as the American Society for Testing and Materials (ASTM) and the American Society of Safety Engineers (ASSE). These and other consensus standards-developing organizations bring industry professionals together to develop national consensus standards under the ANSI process, which helps to ensure representation from a variety of perspectives and interests. OSHA offers cooperation and assistance to these voluntary consensus standards efforts through technical assistance and resources, participation in selected standards development efforts, and sharing occupational safety and health research reports with ANSI to assist in its mission.

Some of the industry consensus standards that are used by employers to supplement OSHA Fall Protection Standards today include ANSI Z359 on fall protection equipment, ANSI A10 on safety requirements for construction and demolition operations, and ASTM E 2505 on rope access. These consensus standards are

[4]OMB Circular A-119 (63 FR 8545 [February 10, 1998]).

generally consistent with OSHA standards, but go into much greater technical detail regarding how to accomplish the requirements set forth by OSHA.

As an example, OSHA sets requirements for rescue after a fall. Specifically, OSHA's Occupational Safety and Health Standards in 29CFR 1910.66 App C Sect I (e) (8) Subpart: F, Personal Fall Arrest Systems states, "The employer shall provide for prompt rescue of employees in the event of a fall or shall assure the self-rescue capability of employees." To help employers achieve OSHA's mandate, ANSI Z359.2–2007 discusses the incorporation of rescue into the comprehensive managed fall protection program; ANSI Z359.4–2007 specifies performance requirements for assisted-rescue and self-rescue systems, subsystems and components; and ASTM E 2505–07 identifies partner rescue as a key part of a rope access program.

Employers must do research on current applicable regulations and standards to ensure compliance for their specific worker type and location. When in doubt it is best to err on the conservative side of interpretation. Employees must be protected not just from falling off a surface, but from falling through holes and from having objects fall onto them from above.

FALL-PROTECTION PLAN

Every employer who places workers at height should do so only under a written fall-protection plan. The fall-protection plan is a thorough treatment of fall hazards, and should not be confused with the job hazard analysis (JHA). A sample JHA form is shown in Figure 1.2.

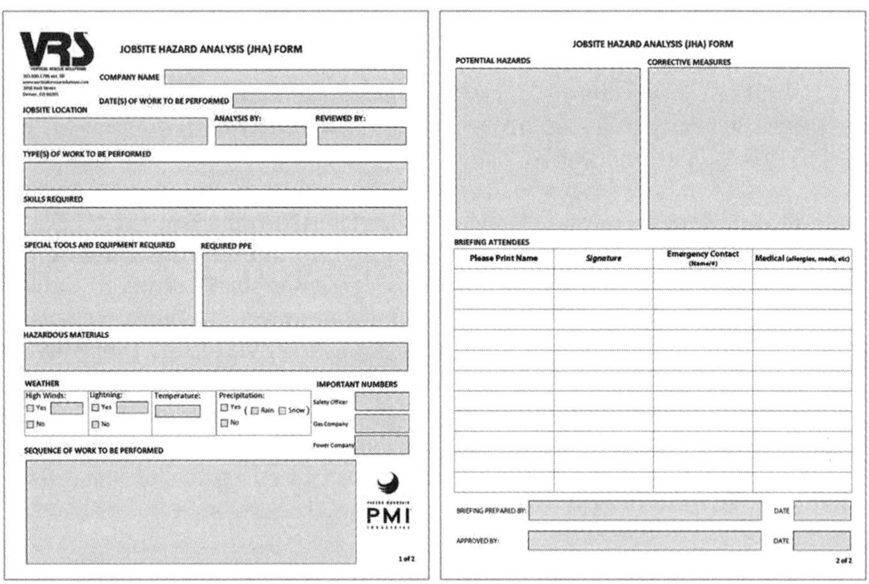

VRS
JOBSITE HAZARD ANALYSIS (JHA) FORM
COMPANY NAME
DATE(S) OF WORK TO BE PERFORMED
JOBSITE LOCATION
ANALYSIS BY:
REVIEWED BY:
TYPE(S) OF WORK TO BE PERFORMED
SKILLS REQUIRED
SPECIAL TOOLS AND EQUIPMENT REQUIRED
REQUIRED PPE
HAZARDOUS MATERIALS
WEATHER
High Winds: Yes No
Lightning: Yes No
Temperature:
Precipitation: Yes (Rain Snow) No
IMPORTANT NUMBERS
Safety Officer
Gas Company
Power Company
SEQUENCE OF WORK TO BE PERFORMED
PMI
1 of 2

JOBSITE HAZARD ANALYSIS (JHA) FORM
POTENTIAL HAZARDS
CORRECTIVE MEASURES
BRIEFING ATTENDEES

Please Print Name	Signature	Emergency Contact (Name/#)	Medical (allergies, meds, etc)

BRIEFING PREPARED BY: DATE
APPROVED BY: DATE
2 of 2

Figure 1.2 Sample JHA Form. Credit: Vertical Rescue Solutions/PMI.

A JHA provides a general overview of hazards while the fall protection plan goes into a thorough, detailed process that is intended to address all aspects of the work to be performed at height. The JHA, then, is but one part of a good fall protection plan. ANSI Z359.2 provides minimum requirements for a comprehensive managed fall protection program and is an excellent resource that provides formal guidance for developing a fall protection program.

A well-written fall-protection plan begins with a policy statement that is specific to the company writing it. This statement provides an overview of general goals and guidance for the managed fall-protection program and emphasizes management's commitment to providing a safe workplace for employees who are exposed to fall hazards. A policy statement might read something like the following:

> *(COMPANY NAME) strives to protect the health and safety of all employees. As a part of this policy, (COMPANY NAME) will equip, train, and expect employees to use effective fall-protection systems whenever exposed to a foreseeable fall hazard. (COMPANY NAME) also maintains an active program to make available self-rescue, assisted rescue, and professional rescue in the event of a fall.*

The fall-protection plan should go on to identify the employer's designated program administrator, as well as the person(s) involved in developing the fall-protection and rescue plans, respectively. The plan should be reviewed periodically, at a minimum whenever work practices, equipment, or other aspects of the program change.

FALL PROTECTION PLAN

A Fall-Protection Plan should include at least the following:

- Policy statement
- Name of program administrator
- Name(s) of responsible person(s)
- Fall hazard survey
- Fall mitigation plan
- Post-fall rescue plan.

The foundation of the plan is the fall hazard survey, which is a thorough, methodical identification of all fall hazards associated with an employer's workplace/practices. The fall hazards should be listed specifically and individually and described in as much detail as possible to facilitate safety and rescue planning. In some cases, a diagram or photograph of the location may be helpful.

In describing the location, be sure to consider how the location is typically accessed, the type of work that normally occurs in that location, persons potentially exposed (not just workers but also bystanders), height of potential fall, obstructions, and other potential hazards. It may also be helpful to consider any history of previous incidents in this (or similar) locations (if known), as this kind of information

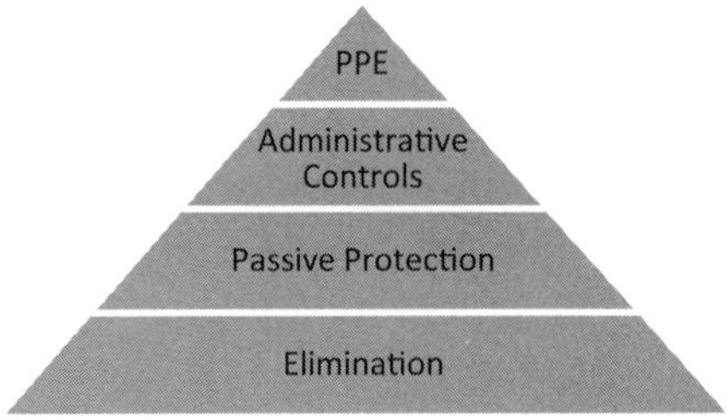

Figure 1.3 Hierarchy of protection.

can provide insight toward prevention as well as provision of safety and rescue. Understanding how likely an incident might be to occur, and how severe the potential consequences might be, will also help drive the response.

The next step is to determine how to best protect employees from each hazard identified. Some people automatically equate the term fall protection with fall arrest, but this is not accurate. Fall arrest is but one of many possible fall protection solutions. Protecting a worker from a fall may be accomplished by any of a number of methods, and employers should always opt for the safest practical means available. The "hierarchy of fall protection," illustrated in Figure 1.3, is a useful tool in determining order of preference. Using this approach, fall-protection methods are presented in an order that is presumed to preferentially provide the greatest safety, with the most desirable method forming the base of the pyramid and the least desirable method at the peak.

As illustrated by the hierarchy diagram, the safest means of fall protection is, of course, to simply eliminate the hazard. Colloquially called "elimination or substitution," in essence it simply means removing the hazard or hazardous work practices. In performing a given task, the employer should ask him- or herself, "Is there is another, safer, practical method that effectively accomplishes the task without placing an employee at risk of a fall?" If the answer is yes, the less risky method should be chosen.

Such elimination is arguably the most effective control measure for protecting employees from a fall in that this prevents the worker from ever being placed in the hazardous situation to begin with. Elimination of a fall hazard is usually achieved by changing a process, sequence, or procedure such that Authorized Persons are not exposed to the fall hazard. Lowering the work surface to ground level—as in lowering highway lights to the ground in order to change the light bulbs—is one example of substituting a work method in order to eliminate a fall hazard. In other words, *fall prevention* is a very effective method of fall protection.

Passive fall protection, the next preferred approach to fall protection, is another preferred method of fall prevention. The advantage of passive systems is that they generally do not require any special training or equipment on the part of the user. Passive fall protection involves isolating the worker from the hazard—for example, by means of a guardrail.

If a hazard cannot be mitigated through elimination or through passive methods, *Administrative controls* may be a good alternative. Administrative controls include processes that prevent a person's exposure to a fall hazard. Such controls might include training, warning signs, lights or sounds, or other warning methods. American National Standard Z535.2, Environmental and Facility Safety Signs, contains standards for warning signs. One disadvantage of administrative controls is that they rely heavily on human behavior, management, and training. Once a person steps over the line in the example above, the immediate consequence is that he or she is exposed to the fall hazard. Under such a circumstance, if the worker *does* fall, fatal injury is likely.

The final approach to protection that should be considered involves use of Personal Protective Equipment (PPE). PPE may be the best solution available when a person must work at or near an edge, or in the vertical environment. Types of protection that utilize PPE include rope access, restraint/positioning, and fall arrest.

Rope access provides excellent protection, but requires specialized training and techniques for workers. Rope access methods utilize a redundant, two-rope system wherein one rope provides the "working surface" and the other provides "backup safety." In effect, it is like a positioning system backed up by a unique, secondary fall protection system—essentially resulting in dual protection. As a result, rope access facilitates safer access to the most challenging of worksites. Using specialized descending and ascending equipment, the worker can move from place to place in suspension to perform a wide variety of work. Rope access workers offer additional advantages in that they are usually highly trained above and beyond work procedures to also effect a self-rescue or partner rescue should the need arise.

Fall restraint is a different method, and involves securing (or tethering) the Authorized Person to a suitable anchorage using a lanyard short enough to prevent the person's center of gravity from reaching the fall hazard.

If rope access or fall restraint methods are not viable options, placing the worker at height and protecting him or her with a *fall arrest* system may be used. Fall arrest is a broad term encompassing a variety of systems designed to stop an Authorized Person after a fall has begun. Fall arrest systems range from fall arrest lanyards with force absorbers to self-retracting lifelines to vertical or horizontal lifelines.

The next few pages will discuss each of these types of fall protection in greater detail, with the goal of providing information that will improve the safety of the rescuer, as well as to his or her ability to render assistance to another.

FALL PROTECTION SYSTEMS FOR RESCUERS

Anyone who may find it necessary to rescue another from an industrial fall protection environment should first have a firm understanding of the various types of fall protection systems used by workers who may require rescue. Would-be rescuers should have a working knowledge of fall protection methods and equipment for their own safety, as well as for greater understanding of how best to achieve rescue in that environment.

The best method(s) of protection for the rescuer's safety in a given environment will most likely be the method(s) already in use there. A firm understanding of fall protection methods and systems will also enable the rescuer to be more effective when rescuing another. What type of fall protection a fallen worker was using will have some impact on how the rescuer(s) are able to locate, access, and extricate him or her.

Passive Fall Protection

Clearly we have determined that workers who are exposed to a potential fall from height while working on a raised surface must be protected from falling by some means. Elevated work surfaces where work is frequent or ongoing may have a permanent or temporary fixed barrier installed around the edges of the platform and surrounding any open holes. These methods of protection are less likely to result in a worker being suspended and in need of rescue than restraint or fall arrest systems. For example, if a worker is not stopped by a railing, chances are they will impact the ground below and rescue will be a relatively simple matter of picking them up from the floor or ground.

This book will endeavor to focus on those types of fall protection systems—both passive and active—which are most likely to stop the worker in a manner or location where the worker may then be in need of rescue. Because barriers and handrails really do not fit this description, this book offers little attention to those methods. However, a few points are notable for rescuers who may be working in environments where such protection is present.

Barriers consist of passive systems such as a guardrail, parapet wall, or fencing. These are intended to provide a barrier to stop a worker from proceeding past the edge of a work level or into a floor opening, and to protect the employee from falling. One type of barrier that is common in many workplaces is the *guardrail*, which consists of a top-rail, mid-rail, and toe-board secured between vertical posts. A compliant guard rail must rise 39–45 inches above the walking/working level and must be capable of resisting at least 200 pounds force at any point along the rails, or if a higher load is anticipated it must be able to withstand that load. There must also be some sort of protection from falling between the rails such as siderails, screens, mesh, or intermediate vertical members.

Stair rails, used as a physical barrier on open stairs and ramps, are built to similar specifications as guardrails. Useful guidance for installation of guardrails, covering holes, and otherwise isolating hazards in general industry may be found in American National Standard A1264.1, Safety Requirements for Workplace Floor and Wall Openings, Stairs and Railing Systems, while similar guidance for construction may be found in A10.18.

Rescuers should note that while most properly constructed guardrails and stair rails offer sufficient strength for their normal, intended use, utilization of a guardrail or stair rail to anchor restraint, positioning, or rescue lines may not be adequate; in fact, some rules specifically prohibit tying off to guardrails.

Rope Access

Rope access is a specialized mode of access and protection that may be employed by specially trained, certified technicians to access hard-to-reach places in a variety of industries. Rope access methods are often combined with conventional fall protection methods (restraint, positioning, arrest) to accomplish engineering inspections, maintenance work, installations, and other tasks in an efficient and comprehensive manner.

Rope access technicians rely on two ropes for access and protection: a primary line for ascending, descending, and traversing, and a backup line for safety. In this type of system, the fall protection offered by the second line is completely separate and independent from the primary means of support. The two rope systems are at once independent yet interchangeable, offering great versatility while maintaining continuous redundancy. Rope access methods are often, but should not be, confused with recreational pursuits such as climbing. This is an erroneous comparison in that recreational climbing is a sport whose focus is the personal challenge of the climbing endeavor, while in rope access, the focus is merely to safely reach a worksite for the purpose of performing a task.

Figure 1.4 shows a typical rope access system. Here you can see that the second line that is employed for backup safety in rope access differs from conventional fall arrest in several ways. First, in order for the secondary system to deploy, the very substantial primary system would somehow have to fail. With built-in fail safe mechanisms and extremely high safety factors, this is highly unlikely. Second, the backup safety system is rigged and managed in such a way as to preclude the extensive fall potential that is allowed in conventional fall arrest. Finally, the backup

Figure 1.4 Rope access system.

safety system is much more dynamic than a conventional fall arrest system, and softens any potential impact with ease.

Rope access technicians are typically evaluated and certified by an independent third party organization such as the Society of Professional Rope Access Technicians (SPRAT). SPRAT is a consensus-driven membership organization that maintains industry-consensus standards on *Safe Practices for Rope Access Work* as well as on *Certification Requirements for Rope Access Technicians*.[5] These useful guidance documents are available from the organization's website at no cost.

Certified entry level rope access technicians typically receive at least 40 hours of initial training, and advanced certification requires extensive experience and verification of greater skill levels. Such extensive training and skills help prevent incidents from occurring, and in the unlikely event of compromise the certified rope access technician is almost always capable of self-rescue and partner rescue.

Fall Restraint

Restraint systems consist primarily of a safety harness or belt, a lifeline and/or lanyard, and a 3,000 pound capacity anchor. These are frequently used where work is being performed in a temporary capacity and/or where a fall arrest system is neither justified nor necessary.

A restraint system performs a little like a leash, and is intended to prevent the worker from reaching an edge where there is a risk of falling. In order to perform effectively, restraint systems must be rigged so that the worker is not capable of reaching a position where a fall might occur.

If the worker misuses the restraint system—for example, if the system is rigged such that the worker can extend his or her body weight beyond a leading edge—it might be possible for a worker to fall into the system and become suspended and in need of rescue. This is clearly a misuse of a restraint system, but planning for such an event may nevertheless be quite justified.

Fall Arrest

Personal fall arrest systems, such as that shown in Figure 1.5, should meet appropriate OSHA guidelines for construction (1926.502(d)) or general industry (1910), depending on the type of use. Normally, a fall arrest system will consist of at least a 5,000 pound capacity anchorage, lanyard, connectors, and a full body harness. Additional components, such as a rope, a fall arrester, and other equipment, may be incorporated into some systems.

A well-designed personal fall arrest system will prevent the user from a free fall greater than 6 feet (1.8 meters) and from contacting any lower level, and will

[5]Society of Professional Rope Access Technicians (SPRAT), 994 Old Eagle School Road, Suite 1019, Wayne, PA, 19087-1866. Http://www.sprat.org.

Figure 1.5 Personal fall arrest system. Credit: Tractel, Inc.

also limit maximum arresting force on the employee to no more than 1,800 pounds force (8 kilonewtons [kN]). At this writing, there is a general move toward reducing the allowable impact force by half, to 900 pounds force (4 kN).

Directly related to the concept of impact force is the matter of stopping distance. When a mass is in free fall and must be arrested, the energy generated by the falling mass has to go somewhere. That *somewhere* often involves force absorbers that extend in length when deployed, rope grabs that intentionally slip against a rope, and equipment that is designed to elongate under load. In fact, considering the allowable free fall distance, an allowable deceleration distance of 3.5 feet, and inherent elongation in the system, it is quite foreseeable that a person could easily end up well in excess of 12 feet beneath the original point from which they fell. This fact is of considerable importance to would-be rescuers.

Effective fall arrest guidelines may be found for general industry in ANSI Z359, or in ANSI A10.32 for construction.

Fall Arrest Using Self-Retracting Lifelines A self-retracting lifeline (SRL) is a specific type of fall protection device that may also be referred to by the monikers "retractable lanyard," "inertia reel," or "fall arrest block." Most SRLs consist of a lifeline (wire rope, webbing, or fiber rope) wrapped several times around a spring-loaded drum that is fitted with a centrifugal brake and encased inside a housing. Some SRLs are designed to be affixed to an anchorage above and in line with the worker, with the end of the lifeline attached to a fall arrest D-ring on his harness. Other SRLs (as shown in Fig. 1.6) feature the block fixed directly to the dorsal D-ring, with the line anchored above. In either case, as the person moves in a

Figure 1.6 One type of SRL. Credit: Reliance Industries, LLC.

controlled fashion the tensioned reel pays out and retracts, allowing the cable to travel but always with a slight bit of tension.

As long as the user moves in a smooth and controlled manner, the drum allows the lanyard to feed out and/or retract, maintaining light tension on the user. In the event of a sudden load, however (greater than 4.5 feet/second), the system locks up with less than 900 pounds force and 24 inches stopping distance. When tension is relieved on the system, the drum once again allows the SRL to move freely.

Fall Arrest Using Vertical Lifelines A vertical lifeline is a rope or cable that is suspended from an overhead anchorage point, as shown in Figure 1.7. Similar in function to the ladder safety system described above, it can be used in a free-hanging configuration, against a structure, on a ladder, with a suspended scaffold, or in any number of ways. A vertical lifeline is not just a device, but is a system, comprised not just of the vertical line itself but also a compatible fall arrest rope grab, connectors, lanyard, and harness.

With the worker connected to the vertical lifeline by means of a rope grab, as shown, the rope grab travels up and down the vertical lifeline as the worker moves in a controlled manner. In the event of a sudden load, such as a fall, the rope grab locks up and catches the fall.

Temporary vertical lifelines may be made of synthetic or wire rope. Permanent systems are usually made of rigid steel or aluminum rails, wire ropes, or similar materials. The fall arrester used must be compatible with the lifeline on which it is used. These can be very line-specific, and misuse can be catastrophic, so this is an important point. Most fall arresters are intended to be used with a force absorber, to help mitigate the impact of a fall. The safety harness is the final piece in the puzzle. This equipment is covered more thoroughly in Chapter 4.

Fall Arrest Using Horizontal Lifelines Horizontal lifeline systems are a versatile solution for work areas where overhead anchorage capabilities are limited, and

Figure 1.7 Vertical lifeline system.
Credit: Reliance Industries, LLC.

may commonly be found on construction sites, roof tops, cranes, rolling stock, and bridges. Essentially, the horizontal lifeline serves as an anchor that facilitates horizontal mobility. The basic elements of a horizontal lifeline system, including anchorage, horizontal line, and PPE, are shown in Figure 1.8.

As in the case of vertical lifelines, the lifeline itself is but one piece of an overall system comprised of harness, lanyard, force absorber, and connectors, in addition to the properly anchored horizontal lifeline. The user connects to the horizontal lifeline by means of a lanyard, which travels across the horizontal lifeline as the user moves back and forth on the structure. Where distances are extensive or complex, multiple horizontal lifelines may be used, as shown in Figure 1.9.

A horizontal lifeline may be used for fall arrest or for restraint. Guidance for the use of horizontal lifelines for fall arrest may be found in 29 CFR 1926.502(d)(8) and there is also some useful information in Appendix C to Subpart M. The most important consideration is that these must be designed, installed, and used under the supervision of a Qualified Person as part of a complete fall arrest system that maintains a safety factor of 2. Because there are so many factors to consider with horizontal lifelines, an engineer often is needed. In the case of horizontal lifelines, it is always a good idea to keep the plan drawings and calculations because OSHA may want to see how the safety factor of 2 was determined.

ANSI 2359.17 (pending) is intended to provide some guidance for their construction. Horizontal lifelines that simply stretch between two points are referred to as single-span horizontal lifelines, while those that pass through intermediate anchorages are referred to as multi-span horizontal lifelines. Multi-span horizontal lifelines

Figure 1.8 Horizontal lifeline system. Credit: Reliance Industries, LLC.

Figure 1.9 Multiple, complementary horizontal lifeline systems. Credit: Reliance Industries, LLC.

may change direction at one or more points in the system. Some types of multi-span systems incorporate brackets and connectors that allow the user to pass intermediate anchors without unclipping, but in most cases workers must use multiple fall arrest lanyards, or twin leg lanyards, in order to effectively pass intermediate anchor points.

Fall Arrest on Ladders Elevated work spaces may be accessed by means of a fixed ladder. Requirements for, and appropriate protection of, a worker who is using a fixed ladder differ from those for a worker who is using a portable ladder, such as an extension or A-frame ladder.

Fixed ladders should ideally be provided with a fall arrest system to protect against a fall. Fall protection on a fixed ladder usually consists of a permanently installed vertical rail or wire rope system that is designed for use with an automatic fall arrest device. As the climber ascends/descends the ladder, the fall arrest device travels freely along the rail or cable. In the event of a fall, the device locks instantly to limit and arrests the worker's fall.

Ladder cages, such as the one depicted in Figure 1.10, may be found on some ladders, but these should not be considered to be a particularly effective method of fall protection.

While ladder cages may provide a means of support to a worker who might wish to lean against the barrier to rest, they cannot be relied upon to actually prevent or arrest a fall. In fact, anecdotal evidence and some research suggest that ladder cages may actually counter the proper function of fall protection systems (such as cables or rails) that are designed for ladders.

Figure 1.10 A ladder cage.

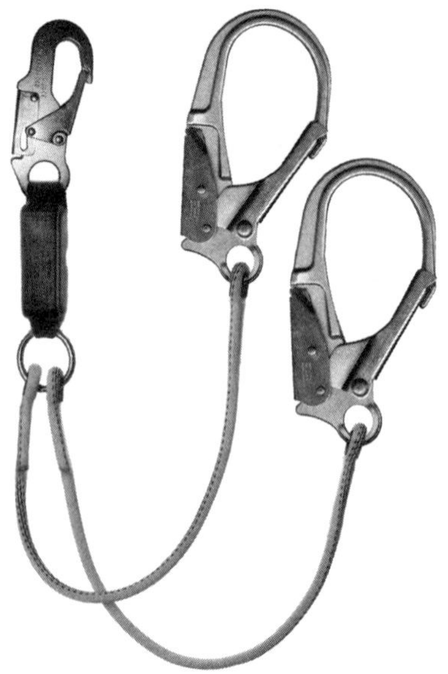

Figure 1.11 Twin lanyards. Credit: The Heightec Group Ltd.

If proper fall protection is not integrated into a fixed ladder design, the climber can use other methods to achieve complete fall protection. For example, the climber who is the first up the ladder may use twin lanyards, as shown in Figure 1.11, to alternately clip into the rails as he or she climbs.

Using this method, as the climber ascends one hook should be placed on one of the side-rails, as high above his or her head as he or she can comfortably reach. While it is also possible to place hooks on the rungs, generally speaking the side rail connection is much more secure and facilitates loading a stronger part of the ladder in case of a fall. When the climber reaches the hook, another hook is placed within reach of the other hand. Once the second hook is connected, the first hook may be removed. This process repeats until the climber has reached his or her goal.

Another possibility is to mount a vertical lifeline or a self-retracting lifeline to a fixed anchor at the top of the ladder. These devices (described in more detail later in this chapter) are utilized along with a full body harness and other components to secure the worker. More information on safety requirements for fixed ladders may be found in ANSI/ALI A14.3.

Fall Arrest on Suspended Work Platforms Where an extensive amount of work must be performed in a relatively inaccessible location, a suspended work platform may be used for employees to gain access. These may be either temporary or permanent in nature, and are commonly found on buildings where frequent access to exterior surfaces, such as windows, is anticipated. Suspended scaffolds (swinging scaffolds) are covered by OSHA in subpart D for general industry and in subpart L for construction, while powered platforms are covered by 1910.66.

Perhaps the most common type of suspended work platform used on buildings is the powered swing stage. Comprised of a work platform, guardrails, and a suspension system, these systems should meet OSHA 1910.66, subpart F. Powered platforms are usually raised, lowered, and leveled by means of one or more motors.

Workers using suspended work platforms should be protected against a fall from height by means of a separate fall arrest system. Although most of these devices feature guardrails, the scaffolds themselves are susceptible to failure due to mechanical malfunction, improper rigging, inadequate anchoring, overturning, collapse, damage, and wind. In these likely scenarios, a guardrail won't be much help. For this reason, fall arrest systems must be anchored independently of the suspended scaffold.

Window cleaners sometimes use a manually controlled miniature version of a suspended scaffold, also known as a boatswain's (bosun's) chair. The concept of the bosun's chair originated at sea. The term boatswain refers to a warrant officer who is responsible for operations on deck. A boatswain's chair, or bosun's chair, historically consisted of a wooden plank with stout lines passing underneath it and up into a bridle at chest level, and was used to haul a sailor up the mast on a halyard to do repair work or inspect the rigging.

Utilization of this tool in circumstances other than sailing was a natural development, especially in coastal regions where people had been exposed to sailing. Window cleaners, chimney sweeps, and even construction workers found a wide variety of uses for bosun's chair-sized suspended scaffolds in the workplace. These systems are typically used and regulated as suspended scaffolds, and should not be confused with rope access techniques.

Bosun's chair systems used in the workplace today have evolved toward improvements in comfort, weight, and strength, and nowadays usually feature a body support strap or straps to help prevent the user from slipping out. Figure 1.12 shows a modern bosun's chair. The movement of the suspended bosun's chair may be

Figure 1.12 Bosun's chair with safety straps. Credit: Tractel, Inc.

Figure 1.13 Rope access technician using a seatboard for comfort.

achieved by another person using a winch or a block and tackle system, or the person in the chair may also be able to raise/lower him- or herself utilizing a mechanical advantage system.

A vertical lifeline system typically protects the worker against a catastrophic fall, but this system does not offer the versatility nor the capability of rope access, nor is the level of training and skill possessed by users of these methods anywhere near that of a rope access technician. Victims of suspended scaffold failures, including bosun's chair failures, are often left dangling from a vertical lifeline by a dorsal attachment, and may be ill-equipped for self-rescue.

Note that the comfort seat used by some rope access technicians may, to the uninitiated, look a lot like a bosun's chair. However, there are some distinct differences between the two. While a bosun's chair really does function as a single point suspended scaffold and serves as the primary point of suspension for a worker, a seatboard is really nothing more than a comfort adjunct and does not serve as the primary source of security for the user (see Fig. 1.13). In a rope access system, the worker's harness is directly connected to the suspension system, so that he or she cannot inadvertently fall out or become detached. The seatboard is not the primary means of suspension, but is attached secondarily only for the purpose of comfort.

Fall Containment Systems

Where it is impractical to provide conventional fall restraint or fall arrest, *containment systems* such as safety nets are sometimes used. Although containment systems may be considered "passive" systems by virtue of the fact that little to no special training is required of the user, they should not be considered to be equivalent of passive systems that *prevent* a fall, such as a handrail. Unlike a handrail, which is intended to prevent a fall from occurring, a fall containment system actually performs more of a *passive fall arrest* function.

Commonly found in bridge work and steel erection projects, safety nets should be installed as close as practicable, but no more than 30 feet, beneath the working surface.

Safety nets are used for a variety of applications, and because they may be as many as 30 feet below the level where work is taking place, rescue scenarios may be fairly complex. As an added challenge, openings in the net may be as large as 6 inches by 6 inches.

ANSI A10.11 provides guidance toward the design, installation, testing, and maintenance of safety nets, and OSHA guidance may be found in 1926.502(c)(1). Industry-specific guidance that is applicable to nets used on water tanks and communication and broadcast towers can also be found in 1926.105. OSHA requires that items that have fallen into safety nets be retrieved as soon as possible, but at least before the next work shift. This means that personnel may be accessing the nets to remove materials, debris, equipment, and tools—on a fairly regular basis.

Aerial Platforms and Man Baskets When there is no working surface on which a person might stand, one solution is for the worker to bring the working surface with them. Aerial baskets, bucket trucks, scissor lifts, and cherry pickers are all examples of mechanical lifts that may be used to temporarily raise a worker to an elevated position for the purpose of performing a task.

Aerial devices are the closest in appearance to a crane, consisting of a number of jointed sections that can be controlled to extend the lift in a number of different directions, often including up and over applications. This versatility permits users to reach locations that can pose particularly difficult rescue scenarios.

Aerial lifts may be powered electrically, or by gas or diesel engines. They may feature articulated arms that allow the basket to be manipulated in any number of directions and are designed with maximum lift limits of up to nearly a ton. Although an aerial lift is often designed to be set up and operated by just one person, this is generally not a good plan. Any work performed at height should always be done in teams of at least two.

Any worker using an aerial platform must be protected against a fall from the platform by means of a fall restraint system or a fall arrest system. While many aerials feature continuous handrails around the platform, they are generally also fitted with an identified anchor point to which a lanyard may be attached. With the safety system attached to the same device on which the worker is standing, catastrophic failure can occur. If the entire platform (or its support) fails or falls, the secondary system will go with it. While unavoidable, this situation does suffer from a lack of redundancy for safety.

Another concern with aerial work platforms is the possibility of the system becoming stuck, or of having to rescue a worker who falls from the work platform.

Some aerial platforms allow the user to bleed off pressure to accommodate rescue in such situations.

Exclusions

In some instances, employers may be able to claim an exclusion from the requirement to provide conventional fall protection. Such exclusions might occur where employees are engaged in leading edge work, precast concrete erection work, or residential construction work, and where it is demonstrated that it is infeasible or it creates a greater hazard to use conventional fall-protection equipment. Exclusions are touched upon throughout OSHA regulations, but the onus is on the employer to justify whatever decision is made.

The safety of the employee should always remain at the forefront of every such decision. Where an exclusion is claimed from conventional fall-protection requirements, the employer is *not* relieved from his or her duty to maintain a safe and healthful workplace. The exclusion simply allows that utilization of OSHA accepted standard practices may not be ideal; still, the employer is required to develop an alternative fall-protection plan that provides other measures to be taken to reduce or eliminate fall hazards for workers.

SUMMARY

Despite the significant effort that has been put into regulation and compliance, falls remain a leading cause of injury and death in the workplace. As we work toward mitigating these hazards, it is imperative to also maintain an ability to respond when an incident does occur.

Regulation in most countries mandates fall protection, and prompt rescue is also required. Some voluntary consensus standards seek to provide guidance and assistance in this process, but there is no substitute for formulating your own comprehensive managed fall protection plan that is specific to your workplace, and for specifying rescue personnel, methods, and equipment as part of that plan.

The remainder of this text provides guidance toward that end.

POST-CHAPTER QUIZ

Chapter 1

1. What is the name of the governmental agency that oversees occupational safety and health in your country?

a) Health and Safety Executive

b) National Occupational Health and Safety Commission

c) Occupational Safety and Health Administration

d) Other: ____________________

2. Which of the following is a *primary goal* of government involvement in occupational health and safety?

   a) Fining employers for misconduct

   b) Ensuring that employers provide a safe workplace

   c) Reviewing employers' personal protective equipment

   d) Developing state plans

3. Match the definitions for the following terms as they are used in this text:

Qualified Person	One who is capable of identifying existing and predictable hazards in the surroundings or working conditions that are unsanitary, hazardous, or dangerous to employees, and who has authorization to take prompt corrective measures to eliminate them.
Authorized Person	One who, by possession of a recognized degree, certificate, or professional standing, and/or who by extensive knowledge, training, and experience, has successfully demonstrated the ability to solve or resolve problems relating to the subject matter, the work, or the project.
Competent Person	An individual who is approved or assigned by the employer to perform a specific type of duty or duties or to be at a specific location or locations at the jobsite.

4. Use words from the box on the right to fill in the blanks in the following diagram, starting with the most preferred method of fall protection at the bottom and the least preferred at the top.

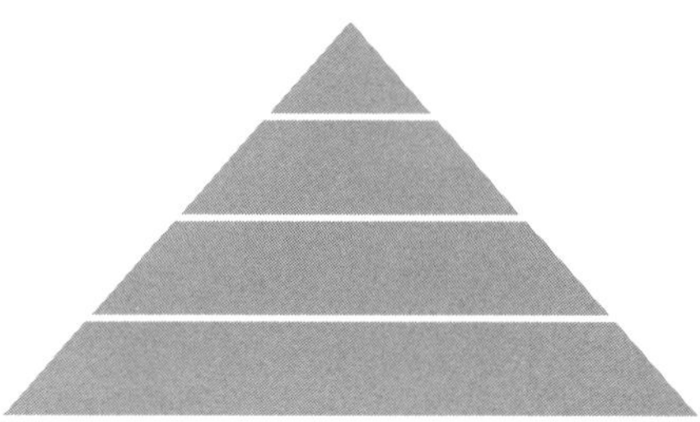

Passive Protection
Administrative Controls
PPE
Elimination

5. True or False: The only way to effectively provide fall protection is through fall arrest.

6. True or False: Ladder cages can be relied upon to provide 100% fall protection.
7. A fall protection plan should contain the following parts (choose all that apply).
   a) Policy statement
   b) Name of program administrator
   c) Name(s) of responsible person(s)
   d) Fall hazard survey
   e) Fall mitigation plan
   f) Post-fall rescue plan
8. A horizontal lifeline may be used for
   a) Fall arrest only
   b) Fall restraint only
   c) Fall arrest or fall restraint
   d) None of the above
9. Guidelines for a comprehensive managed fall protection plan may be found in
   a) ANSI A14.3
   b) ANSI A10.48
   c) ANSI Z359.2
   d) ANSI Z359.4
10. Safety requirements for fixed ladders may be found in
    a) ANSI A10.48
    b) ANSI Z359.2
    c) ANSI Z359.4
    d) ANSI A14.3

ANSWER KEY

1. Answer may vary
2. b
3. **Qualified Person:** One who, by possession of a recognized degree, certificate, or professional standing, and/or who by extensive knowledge, training, and experience, has successfully demonstrated the ability to solve or resolve problems relating to the subject matter, the work, or the project.

   Authorized Person: An individual who is approved or assigned by the employer to perform a specific type of duty or duties or to be at a specific location or locations at the jobsite.

   Competent Person: One who is capable of identifying existing and predictable hazards in the surroundings or working conditions that are unsanitary, hazardous, or dangerous to employees, and who has authorization to take prompt corrective measures to eliminate them.

4. 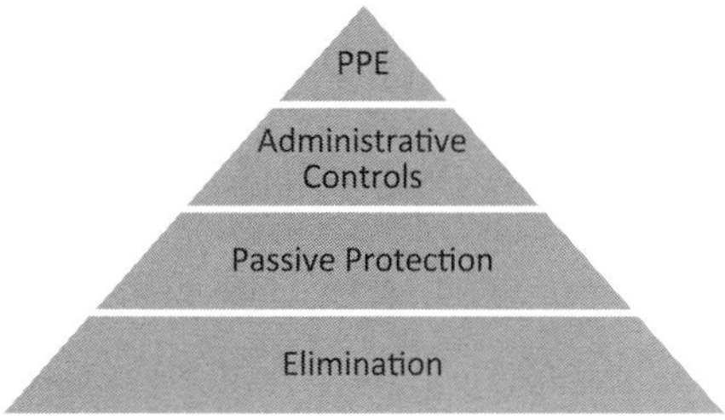

5. False
6. False
7. a, b, c, d, e, f
8. c
9. c
10. d

CHAPTER 2

DEVELOPING A POST-FALL RESCUE PLAN

By the end of this chapter, you should expect to:

- Know when a rescue plan is required.
- Understand other hazards that may exist in a post-fall rescue situation.
- Explain suspension intolerance.
- Explain who the "subject" of a rescue is.
- Identify at least five components of a post-fall rescue pre-plan.
- Explain the *golden rule* of rescue.
- List several important things to consider in establishing a rescue response group.
- List several important performance capabilities to evaluate in a rescue response group.
- Understand how to coordinate with external resources.

In Chapter 1 you encountered some of the regulatory and practical reasons for establishing a post-fall rescue plan for your workplace. But just what *is* a post-fall rescue plan?

This chapter builds on the information in Chapter 1 to help answer that question. It offers insight into the nuts and bolts of what a post-fall rescue plan should cover. While this preplan is essential to the remainder of the book, the reader need not fully comprehend this chapter in order to benefit from other chapters. This chapter is essential for employers, safety managers, and rescue leaders.

SOMETIMES WORKERS FALL

Despite the best of intentions and even with good planning, accidents still sometimes happen. According to the U.S. Bureau of Labor Statistics, falls have been among the top three leading causes of death in the workplace since 1992 and account for nearly 20% of all non-fatal occupational injuries requiring days away from work.

Falls from Height: A Guide to Rescue Planning, First Edition. Loui McCurley.

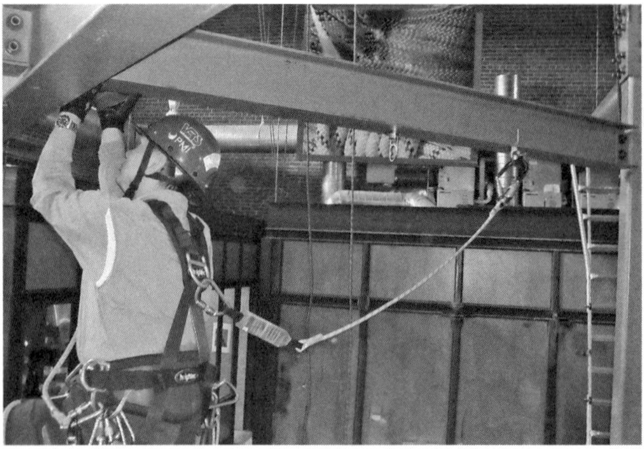

Figure 2.1 A swing fall occurs when there is potential for pendulum.

Catastrophic falls from height in the workplace are not limited to the most obvious or extreme circumstances. Consider the 49-year-old merchandise specialist who fractured his spine while organizing shelves.[1] He was using an order-picking forklift to access storage racks filled with televisions. The lift was equipped with a safety belt and self-retracting-lifeline safety system, but the worker wasn't using it. As the worker pulled two stacked television boxes onto the platform, he suddenly lost his balance and fell backwards, striking a storage rack behind him. In this instant the worker suddenly became what we will refer to in this book as the *subject*—that is, a person who is now the focus of a rescue. Although the distance of the fall was less than 6 feet, this subject's life was forever changed as he sustained fractures to the right side transverse processes of his first four lumbar vertebrae.

When a worker does fall, there are many things to consider. At the forefront of most people's minds immediately after a fall will be the matter of safely reaching and retrieving the person who has fallen. Rescuing the subject efficiently and effectively is important, but not at the risk of injuring others. An effective preplan for rescue will help prevent other workers from reacting inappropriately or causing further harm.

To exemplify the need for prompt rescue consider the maintenance employee who sustained a simple fall when his self-retracting fall protection system was not positioned directly overhead. The offset anchorage resulted in a *swing fall*, slamming him into a steel column.[2] Swing falls, as shown in Figure 2.1, can be quite hazardous. Although the system rigging stopped his fall in only 7 feet, his impact with the column caused a concussion, fractured jaw, cut on right eyebrow, bruised face,

[1]Occupational Safety and Health Administration, Department of Labor, *Fatality and Catastrophe Investigation Summaries, Inspection: 313382004*, http://www.osha.gov/pls/imis/accidentsearch.accident_detail?id=202456786.

[2]Occupational Safety and Health Administration, Department of Labor, *Fatality and Catastrophe Investigation Summaries, Inspection: 312687569*, http://www.osha.gov/pls/imis/accidentsearch.accident_detail?id=201185006.

and bruised right elbow. Prompt rescue was integral to this man's survival and recovery.

In another case, a derrick man was jerked off a platform while tripping a pipe.[3] Although his fall was successfully arrested by a full body harness and lanyard, the harness was not properly fitted and he was strangled to death before he could be rescued.

A fallen worker who remains suspended in fall protection is in greatest danger when he is completely incapacitated, unable to move, or threatened by other environmental dangers. Even a worker who may have been uninjured by the fall itself can become an emergent case if not rescued quickly. There is nothing to compare with the feeling of helplessness and isolation that a worker feels after a fall, and that feeling can quickly translate to a life threatening situation if a suspended worker is not trained and equipped to react appropriately.

SUSPENSION INTOLERANCE

In recent years a lot of attention has been given to the topic of *suspension intolerance*, also called *suspension trauma*, and a sense of urgency has sprung up around the idea of quickly rescuing any worker who is suspended in a harness. While the concern does have foundation in fact, it is important to recognize that it is the motionless suspension (see Fig. 2.2), not specifically the harness, that presents the crux of the hazard.

The term suspension intolerance, then, describes a physiological shock response that is triggered when a person is suspended motionless for a period of time, resulting in the pooling of blood in the lower extremities. When this condition was first recognized, the term suspension trauma was often used to describe it. The term "trauma" may or may not paint an accurate picture, however. The exact physiology of suspension intolerance is debatable, and the medical community is divided on whether it stems from orthostatic intolerance or something else. In fact, the mechanism of injury may vary from incident to incident depending on the progression of events leading up to the suspension, the physiology of the person involved, and whether the person is suspended in a motionless condition (as opposed to being able to wiggle and move their legs to relieve pressure).

Anyone suspended in a harness is potentially subject to experiencing the effects of suspension intolerance, but hanging upright and motionless from an attachment point, especially the dorsal attachment point, appears to be a key precipitator. Other factors, such as weight, overall physical fitness, harness fit, fall distance, and other medical history, may affect a person's ability to assist themselves, but in and of themselves do not seem to have a direct corollary to whether a person experiences the signs and symptoms of suspension intolerance. Time is of the essence—especially if the subject is suspended from the dorsal attachment point. Anyone who may harbor

[3]Occupational Safety and Health Administration, Department of Labor, *Fatality and Catastrophe Investigation Summaries, Inspection: 123273278,* http://www.osha.gov/pls/imis/establishment.inspection_detail?id=307818229.

Figure 2.2 Motionless suspension can be hazardous.

doubts regarding this fact will be easily convinced by temporarily placing themselves in such a suspended condition in a safe, controlled environment, as suggested in Chapter 1[4].

According to Dr. Roger Mortimer, a physician who has studied this condition, what we do know about this condition is that it:[5]

- Is a state of shock from blood accumulating in the legs from *passive suspension.*
- Can result in early fainting with death.
- Can result in late muscle damage.

We also know that a suspended person who is conscious and who can move his or her hands and feet has a distinct advantage over an unconscious subject.

All the quibbles and medical jargon aside, it is generally agreed that a person who is *suspended motionless*, particularly after a significant fall and/or in a compromised environment, is in imminent danger, and this represents a true emergency.

Suspension intolerance has been observed in various harness-related testing projects over the years, but in-depth research was not performed until the 1980s

[4]*Caution!* Anyone attempting this should do so only with a doctor's approval, and with prior preparation for being released immediately upon any discomfort.

[5]R.B. Mortimer, "Risks and Management of Prolonged Suspension in an Alpine Harness," *Wilderness and Environmental Medicine*, 22(1):77–86 (2011).

when the Fédération Française de Spéléologie were prompted to uncover clues as to the cause of death in apparently healthy, otherwise uninjured, suspended cavers who had expired while awaiting rescue.[6] These tests quickly revealed a clear correlation between being suspended without motion and a severe physical response, but the project was discontinued due to the perceived risks to test subjects. More recently, the National Institute for Occupational Safety and Health (NIOSH) in the United States performed tests related to harness suspension,[7] confirming the findings of previous studies and emphasizing that keeping the legs elevated and moving is a key to survival.

As with any medical condition, an ounce of prevention is worth a pound of cure. It is recommended that motionless suspension in a harness be avoided if at all possible. The individual affected by suspension intolerance may experience signs such as increased difficulty in breathing while suspended, a growing discomfort in extremities, tingling or numbness in the legs, and may eventually lose consciousness. Personal escape becomes increasingly difficult as the condition progresses.

A worker who has fallen and is suspended in his or her harness should shift and wiggle in the harness and try to keep his or her legs moving or even elevated. Of course, on a practical level, it is quite difficult to keep one's legs elevated and to keep moving after a fall, particularly if one is hanging from a dorsal attachment point in a fall protection harness. If the structure is accessible, the fallen worker may be able to use it to push against or step up on. Otherwise, the worker may use a foot loop or other device to step up into, as illustrated in Figure 2.3, to help relieve pressure from the groin area. In any case, the real take-home message here is that any person suspended after a fall during work at height needs to be rescued promptly so as to avoid becoming motionless, unconscious, physically compromised, and/or debilitated by the effects of this condition.

Prompt rescue should be provided to anyone who is suspended in a harness, particularly if the person is unconscious or unable to move. Rescuers responding to a person who is experiencing suspension intolerance may observe symptoms including swelling of the face and hands, puffiness around the lips and eyes, shallow breathing, and extreme lividity. On the other hand, responders may not notice any observable symptoms at all. An absence of symptoms does not necessarily mean an absence of the condition, and subjects should be closely observed during and after an incident.

Treatment for the condition was once believed to be to keep the subject upright, in a seated position, even after they are on the ground. However, more current protocols suggest that treating for shock as normal, including elevating the legs, high flow oxygen, and fluid replacement, is the preferred approach.

[6]J. Bariod, "Pathologie induie par le harnais," *Commission Médicale Fédération Française de Spélélologie*; J. Bariod, "Report on Experiments with Harnesses," *Medical Commission of the French Federation of Speleologists*; J. Bariod, "Pathology of the Harness for Caving: Additional Dossier to the Report on Experiments Conducted in 1984 at Chalain," *Medical Commission of the French Federation of Speleologists*.

[7]Nina L. Turner, James T. Wassell, Richard Whisler, and Joyce Zwiener, "Suspension Tolerance in a Full-Body Safety Harness, and a Prototype Harness Accessory," *Journal of Occupational and Environmental Hygiene*, 5(4):227–231 (2008).

Figure 2.3 A conscious subject may be able to use a foot loop to relieve pressure.

In addition to rescuing the subject from the fallen condition, the employer must also consider other hazards that may exist within the environment where the employee is working, prepare for treatment of the subject's potential medical condition(s), and maintain safety for everyone involved. Advances in knowledge and equipment have helped to reduce the timeline involved in providing high angle rescue, which is key to avoiding suspension intolerance. Being prepared, having the right equipment, and knowing how to use it is essential to a positive outcome.

CREATING A PREPLAN

Work at height clearly involves situations wherein life threatening accidents are a potential hazard. In order to comply with regulatory requirements and provide appropriate care for employees, an employer should designate a rescue plan, provide appropriate rescue equipment, and train/prepare designated employees accordingly. These capabilities may come from either on-site or off-site resources.

In this manual, we will refer to the process of preparing in this manner by the term *preplan*. While this term may seem redundant (who makes a post-plan?) it is important in emergency response to distinguish between the planning that takes place prior to an incident occurring as compared with that which takes place as part of managing an actual incident. Planning does not end with the preplan, but after an incident occurs the real-time operations planning leans heavily on the preplan. There are two types of planning, then: the preplan and the operational plan. Our focus in this chapter will be the preplan, that which occurs in preparation for an event which has not yet occurred. In Chapter 3 we will address the operational plan.

At this writing only a limited number of OSHA standards specifically require a rescue plan, but proposed subpart I fall protection rules that are pending at this time are expected to specifically require prompt rescue for workers at height regardless of the type of personal fall protection equipment being used. Even where rescue is already mandated,[8] OSHA Fall Protection regulations do not specifically require that a fall rescue preplan be *written*. Nevertheless, a written plan is arguably the best way to develop and monitor procedures, ensure that appropriate equipment is available, and provide consistent and adequate training to affected and involved personnel.

Employers are also required by various OSHA standards[9] (and the general duty clause) to take appropriate precautions to assure that the rescuers themselves do not become victims. Rescuers are just as susceptible to workplace hazards as any other employee, and there is no exemption from OSHA regulations for rescuers. For example, if a worker walks out onto a beam to help a companion who is suspended in fall protection, he or she must protect him- or herself from the potential fall to which he or she is exposed.

It is often (incorrectly) assumed that OSHA will not issue a citation to an employer for an employee who places him- or herself in imminent danger while trying to rescue or assist someone else. This is not necessarily accurate. Although OSHA may be quite forgiving when it comes to spontaneous rescue efforts, and often does not cite for incidents incurred by a co-worker trying to spontaneously rescue a fellow employee, OSHA *may* cite if the employer fails to have a rescue plan in place to begin with, if someone is injured during a rescue attempt, or if the employer fails to provide appropriate training and rescue equipment.[10] These fines can reach thousands of dollars, and as such it is in the employer's financial and ethical interest to be proactive in planning and preparing for safety.

One manufacturing facility was cited for seven willful violations when two employees suffered fatal burn injuries while performing welding operations inside a confined space which had just been painted. One of the violations cited included lack of an established rescue team.[11] In another incident, OSHA proposed $1,610,000 in fines against a grain handling facility for 23 alleged willful violations after the

[8]29CFR 1910.66 App C Sect I (e) (8) Subpart F, Personal Fall Arrest Systems and 1926.502(d)(20) Subpart M, Fall Protection.

[9]Occupational Safety and Health Administration, Department of Labor, 29 CFR 1910.119(n) Process Safety Management. 29 CFR 1910.120(1), (p) and (q) Hazardous Waste Operations. 1910.272(d), (e), and (g) Grain Handling. 29 CFR 1926.106, 1926.651(g) Work Performed Near or Over Water and Excavation Work 29 CFR.

[10]Occupational Safety and Health Administration, Department of Labor, December 27, 1994. 29 CFR 1903.14 Policy Regarding Employee Rescue Activities.

[11]Occupational Safety and Health Administration, Department of Labor News Release, USDL: 95-359, "OSHA Proposes Nearly 0.5 Million Dollars in Fines against Trinity Industries," http://www.osha.gov/pls/oshaweb/owadisp.show_document?p_table=NEWS_RELEASES&p_id=1171.

death of an employee.[12] Among the violations cited were failing to provide observers equipped to provide assistance, failing to train workers, a lack of rescue equipment, and failing to implement an emergency plan prior to entry.

In OSHA regulation 1926.503(a)(2)(vii), employees are required to receive training that addresses their respective role(s) and responsibility(s) within a fall protection plan. Preparing to respond to a fall is clearly an important aspect of an adequate fall protection plan. For an employer to effectively meet the rescue requirements of the fall protection plan, some preparations must be made in advance.

A number of OSHA-published compliance policies and interpretations provide specific directives related to emergency response.[13] These can be an excellent starting point, but nothing can take the place of practical, hands-on training, and holding the occasional drill to test whatever system the employer has established for responding to an emergency. This is the heart of the matter, and will be the emphasis of this text: proper preparation and training of personnel to undertake rescue in the high angle environment.

EMERGENCY RESPONSE PLANNING

Most employers are familiar with the concept of using a written *emergency action plan (EAP)* to facilitate and organize employer and employee actions during workplace emergencies. Exactly which types of emergencies are covered in an EAP will vary from one workplace to another, depending on what environments people may find themselves working in and the hazards that might be most likely to occur.

There are federal regulations promulgated by OSHA that provide specific guidance for employee emergency plans. These regulations provide a useful overview for preplanning for emergencies, and resources exist for both general industry and construction.

According to OSHA standards, an EAP is a written document that must include but is not limited to:[14]

1. Means of reporting fires and other emergencies.
2. Escape procedures and personal escape route assignments.
3. Procedures to be followed by employees who remain onsite to conduct critical operations before they evacuate.

[12]Occupational Safety and Health Administration, Department of Labor News Release. 10-711-DEN, May 27, 2010. "US Department of Labor Fines South Dakota Wheat Growers Association of Aberdeen, SD, More Than $1.6 Million for Grain Handling Violations Worker Suffocated in December 2009 after Being Engulfed in Grain," http://www.osha.gov/pls/oshaweb/owadisp.show_document?p_table=NEWS_RELEASES&p_id=17805.

[13]Occupational Safety and Health Administration, Department of Labor. 29 CFR 1910.146 Confined Spaces in General Industry; 1910.66(e)(9) Powered Platforms for Building Maintenance; 1926.502 Fall Protection Systems Criteria and Practices; Interpretation Letters, Michael Wright (December 18, 2003), J. Nigel Ellis (May 11, 1999), and Charles Hill (August 14, 2000).

[14]Occupational Safety and Health Administration, Department of Labor, 1910.38 Employee Emergency Plans in General Industry; 1926.20, .21, and .35 Training and Emergency Action Plans in Construction.

4. Rescue and medical duties for those employees who are to perform them.
5. Procedures to account for all employees after emergency evacuation has been completed.
6. Names and/or job titles of persons who can be contacted for further information or explanation of duties under the plan.

These requirements involve the big picture of protecting workers and being prepared to respond to common workplace emergencies. Generally, post-fall rescue is not covered by EAP protocol; most EAPs simply address the most obvious and common natural and other potential disasters such as flood, fire, tornado, and hurricane. Although EAPs do not typically address post-fall rescue, a potential for falls deserves no less consideration. Where potential hazards involve a possible fall from height or confined space incident, EAP guidelines provide an excellent framework for planning.

In addition to general requirements set forth in the Occupational Health and Safety Act and overview statements, OSHA addresses emergency response specifically in context of medical care, hazardous waste operations and emergency response (HAZWOPER),[15] respiratory regulations, and fall protection.[16]

Dialing 911 to activate local emergency services is not, in and of itself, considered to be an appropriate "rescue plan." However, advance coordination with local emergency services providers to identify what capabilities they may or may not be able to provide is an important part of any rescue plan, particularly for worksites where employees may be exposed to hazards other than the usual "office-type" risks.

Employers should ensure that they have information regarding local municipal responder resources on hand, including hospital, ambulance, fire, and other services. Address information, direct-contact telephone numbers, and the name of a personal contact with each agency are useful resources to include in the planning process and to have on hand in case of emergency.

PREPARING FOR POST-FALL RESCUE

Rescue after a fall is something that any employer with workers at height should consider. The *post-fall rescue plan* should identify what immediate action(s) a worker should take to notify co-workers, company representatives, or other authorities in the event of a fall, and what next steps should be taken and by whom. The post-fall rescue plan should also provide some insight as to the minimum and maximum amount of time it will take for responders from municipal agencies to reach the suspended worker, as well as what capabilities they might have when they arrive.

Employers should contact local emergency services organizations in advance to discuss and ensure a timely response and proper preparedness upon arrival, especially in workplaces where there is high probability of a worker being exposed to a potential fall. Not all municipal response agencies have ready access to high angle

[15]29 CFR 1910.120(1), (p) and (q) Hazardous Waste Operations.

[16]1926.502 Fall Protection Systems Criteria and Practices.

rope rescue capabilities, and if response to this type of incident requires a separate call out it would be best to indicate that from the outset rather than waiting until the first responders get there and then having to send out a second alarm. It is also worth putting reasonable effort into preparing local responders for what they will encounter when they arrive. Allowing response organizations to conduct annual training on site, and providing tours for them any time exposures change, can significantly reduce the response timeline in the event of a true high angle emergency.

Sometimes it is simply not feasible for municipal emergency responders to reach a potential fall injury victim within a reasonable amount of time (that is, a short enough time to prevent further harm). In this case, the employer should establish an alternative means of assisting, securing, and removing the subject from harm.

When it comes to post-fall rescue, OSHA doesn't provide many detailed requirements; this is at least in part because OSHA does not have a crystal ball to see every possible circumstance, hazard, or work situation. However, OSHA does mandate in some sections of regulation that employers make provision for prompt self-rescue or rescue of employees in the event of a fall.

OSHA REQUIREMENTS FOR RESCUE AFTER A FALL

29CFR 1910.66

App C Sect I (e) (8)

Occupational Safety and Health Standards

Subpart F, Personal Fall Arrest Systems

The employer shall provide for prompt rescue of employees in the event of a fall or shall assure the self-rescue capability of employees.

1926.502(d)(20)

Safety and Health Regulations for Construction

Subpart M, Fall Protection

The employer shall provide for prompt rescue of employees in the event of a fall or shall assure that employees are able to rescue themselves.

While OSHA language presents *self-rescue* versus *prompt rescue* as an either/or option, prudent employers will prepare for both self-rescue and prompt rescue by others. While self-rescue is certainly an admirable, preferred goal, self-rescue capability can never really be *assured*. For example, a worker may be incapacitated by an injury, illness, or even shock, rendering him or her incapable of self-rescue. There is simply no way to completely assure that a worker will not become incapacitated. The wise employer, then, will assure that anyone working at height is capable of assisting him- or herself, but will also have a fast-acting backup plan should rescue by others become necessary.

It is important to quantify the term "rescue" here. Although the term self-rescue has come to refer to the act of recovering oneself from an incident, one who is trained merely to manage oneself in an emergency can hardly be considered a "rescuer." The term "rescue" more accurately suggests to the listener a capability of removing *another* person from harm. Any person who is authorized to work at height should be capable of using available means and methods to extricate him- or herself from a hazardous situation, but it would be erroneous to refer to a person who is so trained as a "rescuer" or as being "trained in rescue." The person would merely be trained in self-care when working at height. In order to avoid confusion throughout this manual, we will use the term "personal escape" to refer to the concept of self-rescue, whereas when the term "rescue" is used it will be to denote the concept of rendering assistance to another.

In this book we will recommend a progressive approach to responding to a fall, beginning with the idea that *all* authorized at-height workers should be capable of personal escape—that is, escaping from the predicament in the event of a fall—and that in most cases it is advisable to also have an plan for immediate, onsite response, usually provided through *co-worker assisted rescue*. Finally, we will also recommend that employers make advance preparation for an expanded capability, which may be provided by offsite resources—most often incorporating *professional rescue* capabilities from local emergency services responders.

In terms of rescue capability, it is not enough to simply have the right equipment on hand, nor is it sufficient to have participated in a class or two on the subject. True *rescue capability* requires a coordinated combination of the right equipment and the proficiency to use it.

Wherever work at height is taking place, the corresponding rescue plan should include provision for

- Personal escape
- Co-worker assisted rescue, and
- Professional rescue.

In addition to the at-height considerations, some workplaces may involve additional potential for hazards, many of which are addressed by OSHA through specific standards. Some of the industrial processes and environments for which OSHA provides more detailed rescue related precautions and safeguards include confined space, hazardous waste, and work over water.

OSHA Standards That Address Rescue-Related Operations:

- Process Safety Management 29 CFR 1910.119(n)
- Hazardous Waste Operations 29 CFR 1910.120(1), (p) and (q);
- Confined Spaces in General Industry, 29 CFR 1910.146
- Confined Spaces in Grain Handling, 1910.272(d), (e), and (g)
- Construction Work Near or Over Water, and Excavation 29 CFR 1926.106, 1926.651(g)

Be very clear on these regulations. Authorized Persons (whether working or rescuing) must be able to meet all requirements of all regulations that pertain to a given operation, not just the height-safety portions. As previously discussed, although OSHA has stated that they will not issue a citation to an employer for spontaneous rescue activity, this does not relieve the employer from the obligation to *prepare* for rescue where fall potential exists, and to ensure safety to employees who may be *designated to respond* in the event of an emergency. In other words, although a worksite at which a spontaneous rescue occurs by necessity may not be cited for breaches during the rescue itself, the employer may be cited for failing to have an adequate rescue plan in the first place—even if the rescue in question was successful![17]

Post-Fall Rescue Capabilities

A post-fall rescue plan must be grounded in the golden rule of rescue; that is that *no further harm* will be created by the rescue attempt—either to the subject or rescuers.

It is not necessary for all aspects of post-fall rescue capability to come from an in-house response group, nor is such a practice prohibited. Either of these solutions—or any number of alternative solutions—is acceptable. What is necessary is for any organization employing workers at height to have some means to ensure prompt rescue of any worker at each specific worksite where an employee might fall. A plan for achieving this ability should include, at a minimum, protocols to report the incident, personal escape techniques to be used by the fallen employee, assisted rescue techniques that can be implemented within just a few minutes, expanded technical rescue techniques for more complex scenarios, requirements for communication amongst group members, and medical/first aid provision.

While many people may consider the act of dialing a nearby municipal emergency response organization to be "a rescue plan" of sorts, this by itself is seldom acceptable as a complete solution, either for post-fall rescue or even for simple medical response. Effective rescue requires a prompt response—meaning that assistance should be able to reach the subject quickly enough to prevent further harm. Based on letters of interpretation from OSHA, an elapsed time of 15 minutes[18] from time of incident may be used as a reasonable guide in workplaces where the possibility of serious work-related injuries is relatively remote. On the other hand, in workplaces where serious accidents, such as those involving falls, suffocation, electrocution, or amputation are possible, a timeframe as short as 4 minutes[19] may be deemed more appropriate.

[17]Memorandum for: Regional Administrators. Through: Leo Carey, Director, Office of Field Programs. From: H. Berrien Zettler, Deputy Director, Directorate of Compliance Programs. Subject: Policy Regarding Voluntary Rescue Activities, March 31, 1994, http://www.osha.gov/pls/oshaweb/owadisp.show_document?p_table=INTERPRETATIONS&p_id=21548.

Memorandum for: Regional Administrators. From: James G. Maddux, Director, Directorate of Construction. Subject: Enforcement of Steel Erection, Subpart R, May 21, 2012, http://www.osha.gov/pls/oshaweb/owadisp.show_document?p_table=INTERPRETATIONS&p_id=28181.

[18]Occupational Safety and Health Administration, Department of Labor, "Interpretation Letter," Mr. Bruce Sutherland, February 5, 2007.

[19]Occupational Safety and Health Administration, Department of Labor, "Interpretation Letter," Mr. James Stewart, April 27, 2004.

Developing and evaluating a rescue service for post-fall rescue can be a daunting task. A good place to start is with a review of the fall protection plan for the jobsite. As discussed in Chapter 1, a fall protection plan should be developed as a part of work preparation for every site where work is performed at height, and any fall protection plan should incorporate a job hazard analysis (JHA). This fall protection plan, and its corresponding JHA, should be updated whenever conditions change, so as to adequately ensure safety during work. Because the JHA specifically examines *all* hazards (not just those related to the at-height work), it should not be overlooked as a useful tool in developing the rescue preplan. Employers may also need to add criteria, based on their specific needs, hazards, and environments, to supplement this information. Whether the fall protection plan, JHA, and post-fall rescue plan are each individual documents, or are combined together in a single document, is up to the employer. The important thing is merely that all of these be addressed.

A written plan should be developed to clearly guide response protocols for every location, and every shift, where post-fall rescue may be required. Affected personnel should be adequately trained and prepared in accordance with this plan. Some of the questions the plan should seek to consider are listed below:

1. Which employees, and what part(s) of the worksite, will this plan cover?
2. During what time(s) of day might an incident occur?
3. How many people might be exposed at any given time?
4. In addition to the fall hazard(s), what other potential hazards might exist at this site?
5. What personal escape equipment and techniques will be used by a fallen worker at this site? (Include detailed diagrams, anchorage information, techniques and methodologies, equipment information, etc., as appropriate.)
6. Who will respond to provide *assisted rescue* to the subject of a fall accident occurring at this worksite?
7. How will the *assisted rescue* responder(s) be notified?
8. What *assisted rescue* equipment and techniques will be used to rescue a fallen worker at this site? (Include detailed diagrams, anchorage information, techniques and methodologies, equipment information, etc., as appropriate.)
9. Who will respond to provide *professional rescue* to the subject of a fall accident occurring at this worksite?
10. How will the *professional rescue* responder(s) be notified?
11. What *professional rescue* equipment and techniques will be used to rescue a fallen worker at this site? (Include detailed diagrams, anchorage information, techniques and methodologies, equipment information, etc., as appropriate.)
12. Who will render post-fall medical care to the subject of a fall accident occurring at this worksite?
13. How will the medical care responder(s) be notified?
14. To what medical facility will a fallen person be transported?

Worksite Considerations

The first step in preparing for rescue is to define the area of responsibility that each plan will encompass. A given worksite may develop a single plan for a wide variation of needs, or separate plans for specific areas of need. This will be partially determined by worksite dimensions, although scope of work and other nearby hazards will also play a part in this determination. Worksites that are geographically separated, that are particularly large, or that involve limited access will be more likely to have different response plans for different aspects of the jobsite.

Knowing how remote a worksite is, what kind of work is taking place there, whether the plan covers a subdivision of a larger jobsite, and how many employees might be in the area are examples of important considerations. This kind of information should be available in the fall hazard analysis document discussed in Chapter 1.

Realizing that OSHA interprets "prompt care" as being up to about 15 minutes, response plans should take into consideration how quickly "assisted rescue" can reach any given individual in the area covered by the plan. With this in mind, this text will focus on the concept of developing protocols for assisted rescue *within the workplace.* For simplicity and clarity, the term "co-worker assisted rescue" will be used throughout the text, although it should be understood that those providing such service may or may not actually be "co-workers" in the truest sense of the word. They may, in fact, be subcontractors, report to a different employer, or be part of an internal emergency response organization. The key to co-worker assisted response is that those responding must be capable of reaching the subject within a few short minutes.

Clear identification of the physical boundaries of a response plan will help in planning resources. In determining response time to an incident, consider whether help will be coming from inside the workplace, from an external location, or from a combination of both. If help is coming from inside the workplace, consider the type of work the responding co-workers are likely to be involved in at the time of an incident. As discussed in Chapter 1, persons at risk should never be working alone, but still it takes time for even a nearby co-worker to respond. A responding person who must first get off-rope, ascend several flights of stairs, and negotiate a catwalk using fall protection equipment and techniques in order to reach the subject will have a very different response time than a fellow employee who can simply set down his or her wrench and take an elevator to the level of, or above, the fallen worker.

Finally, consider which employees are covered by the plan in relation to all of the employees who are actually on site. While the intent of the employer might be to cover only his or her own workers, if different levels of contractors or subcontractors are on site, should a problem occur it is unlikely that his or her trained responders will be willing to stand by and wait for help to arrive. It is in the best interest of all employers on a given site to engage in open dialogue regarding this subject, and to have a plan for coordinated response. Chapter 3 will discuss multiemployer worksites in greater detail.

Contributing Factors

Potential hazards, as well as appropriate response protocols, will fluctuate at a given worksite depending upon a variety of factors. A response to an incident that occurs on a Tuesday morning at 10:00 a.m. when a plant is fully staffed might look very different from an incident that occurs at 10:00 p.m. on a Saturday night with a skeleton work crew in place. Not only would such a scenario affect the proximity and availability of those to assist, a nighttime incident might also be hampered by limited visibility, temperature, or other factors.

These variable factors might also contribute to just how hazardous a particular situation might be at any given time, and these too should be noted. For example, environmental conditions (weather, temperature, lightning, high winds), surface contaminants (oils, fluids, chemicals, even water), work tools being used, and the experience level of worker(s) can completely change the outlook of the work itself, as well as any rescue attempt. Consider, for example, an ice cream factory warehouse, where workers are exposed to heights up to 250 feet in temperatures of –40°F. In such a situation, the entire rescue must be completed within 30 minutes in order to avoid exposing the subject and rescuer(s) to further harm from cold-related hazards.

Whenever a worker falls, consideration must be given to why that fall might have occurred in the first place and the would-be rescuer should use this information to ensure that he or she is protected adequately. A simple slip or trip is one thing, but what if the fall was the result of a medical condition? Or what if an environmental condition, such as toxic air, was the cause? Consider the probable mechanism, and be prepared to evaluate the situation adequately before exposing additional employees or responders.

Any employee who may be designated to rescue another must be provided at least with protection equivalent to that which would be appropriate for a worker in the same environment. This includes personal protective equipment, communications, breathing apparatus, and any other appropriate safety equipment.

Protocols

From here, we can begin to examine existing response protocols. To begin with, describe how your organization would respond to each of the hazards identified in the JHA if an incident were to occur today. You may find that some of these responses are appropriate, while in others you may see significant limitations. Examining these limitations is where the rescue preplan begins to really take shape. Using a critical eye, examine the limitations of each of your existing response protocols, making note of where improvement is warranted.

As you examine your existing response protocols, decide which (if any) can be simply adopted into your new emergency response preplan. For those with which you are not satisfied, develop a new preplan. Every hazard identified should be offset by a specific emergency response preplan, ideally all within the same document, but allowing for different combinations of hazards.

The preplan generally should take into consideration such things as electrical hazards, mechanical rotation, confined spaces, the need for *lockout-tagout (LOTO)*, tower access considerations, environmental hazards, and fire. It may be useful to use a checklist approach call out personal escape protocols, assisted rescue protocols, and external response protocols, including methodologies at every level for:

- Notifying/initiating responders
- Individual responsibilities
- Equipment and systems to be used, and
- Criteria and means for activation of the "next level" of response.

These are the meat of your emergency response preplan, and may take some time to develop. You may find that soliciting assistance from other safety and/or rescue professionals during this phase can help to prevent tunnel vision and promote creative solutions. This plan may also change over time, as your response group resources and capabilities evolve.

When developing a fall rescue response plan,

1. Define area of responsibility.
2. Identify known or foreseeable fall hazards. Consider
   a. Number of persons exposed
   b. Contributory factors (environment, task, experience, etc.)
   c. Time to rescue
3. For each hazard, develop a *personal escape* protocol, an *assisted rescue* protocol, and an *external response* protocol, including methodologies at every level for:
   ◦ Notification of incident
   ◦ Individual responsibilities
   ◦ Equipment & systems to be used
   ◦ Criteria and means for activation of the "next level" of response
4. Train, practice, assess, re-train.

Notification of the Incident

The first challenge in responding to an incident lies in knowing that a response is needed (see Fig. 2.4). Somehow someone must find out that a worker has fallen, and then they must activate the plan. This is one reason why many employers require personnel to work in teams of at least two. This is good practice, and allows the workers to look out for one another.

Where workers are not in constant visual or verbal contact, a whistle may be a viable tool to help one notify the other in case of an accident—but of course the subject must be able to access and blow the whistle in order for this to be effective. Another method that is often used is pre-established radio check-in times, wherein

Figure 2.4 Notification of an incident may be by any method that will achieve the desired response.

each worker must verbally check in with a supervisor or central coordination point at periodic intervals. Again, this method requires action on the part of the subject, and clearly has drawbacks in the case of an incapacitated worker. In either of the above methods, an incident may not be noticed immediately.

One option that does not require user-intervention is to equip each worker with a *personal alert safety system (PASS)* device. Particularly common in confined space operations and among professional rescuers entering a hazardous environment, the PASS emits a loud, audible alert when the wearer is inactive for a preset period of time (usually 30 seconds). Most PASS devices can also be activated manually in an emergency, such as when an employee is injured.

The first person(s) to learn of an incident will probably be those who are working nearby. Their first instinct may be to rush to the assistance of the affected worker(s), but doing so without first notifying someone else could be catastrophic. The necessity and the means for notifying the appropriate next levels of response before taking action should be clearly communicated and understood by every employee. At some level of notification, a company-wide alarm may be issued to instruct employees to collect at a predetermined gathering point.

Keep in mind that cell phones don't always work very well in every environment or workplace. It may be necessary to use handheld radios to announce/notify site workers of an emergency condition and to relay to someone who can get outside communications. Handheld radios are also useful for communications between the onsite emergency coordinator and responders during a rescue. High angle workers and those who respond must be trained in how to use the radios and associated communication protocols if they are so equipped.

> For any potential incident, a preplan should be established for the worker to personally escape, for co-worker assisted rescue, and for professional rescue. At every level of response, when a response is initiated the next potential line of response should also be notified in case the situation does not readily resolve.

Individual Responsibilities

When an incident occurs, everyone involved should know what their responsibilities are and should be trained to respond immediately as directed. This includes everyone who is onsite, even if their responsibility is just to relocate to a predetermined gathering point and stay out of the way.

The post-fall rescue plan should clearly identify leadership roles during an incident so that it is clear who is making key decisions. That person, or those persons, should be notified as soon as possible so that they can begin coordinating the effort. Certain individuals within the company who may be designated to provide co-worker assistance should also be among the first to receive and respond to the news.

While the company CEO is generally where the ultimate responsibility lies, the response plan may specifically identify other individuals to manage key functions during an incident. For example, internal operations may be deferred to someone with training or experience in emergency incident management, while communications may be deferred to a public relations executive. Allocation of responsibilities within a workplace response group is discussed in further detail in Chapter 3.

Information management is an important consideration during an incident, both internally and externally. Notification to company executives is of prime importance and of course the family(s) of the affected employee(s) should also be notified by an authorized company representative as soon as practicable. Depending on the severity of the situation, provision may need to be made for relaying information internally in order to prevent overreaction or panic from employees. External communications include plans for handling public relations, media, legal, and other related matters.

Especially in these days of electronic communications and social networking, companies should have clear policies and employees should understand the company's expectations insofar as external communications are concerned. Only authorized individuals should be permitted to make official statements regarding an incident, especially while response to the incident is still under way.

Personal Escape

By now it should be clear that personal escape is the first line of response for any person working at height. Any individuals using fall protection/fall arrest equipment should be equipped and prepared to extricate themselves in the event that they fall

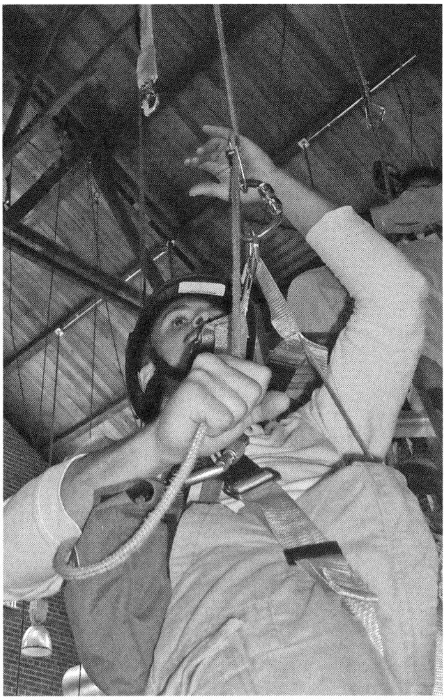

Figure 2.5 Personal escape is the preferred response for a fallen worker.

and are caught by their safety system, as in Figure 2.5. In order to be proficient at such skills during a true emergency, a worker must be very familiar with those skills in non-emergent conditions. Knowing and practicing appropriate personal escape techniques in advance, using equipment at hand, is essential.

Specific skills, equipment, and readiness for personal escape will be discussed in future chapters. Note, however, that any time an employee takes a fall and must utilize personal escape skills, the next level of response capability identified in the preplan should be notified and initiated.

Co-Worker Assisted Rescue

Co-worker assisted rescue involves utilization of limited techniques by personnel who are *onsite* and capable of immediate response, to make contact and to remove a subject from harm within a short period of time. This differs from professional rescue techniques in that the responders may not be professional rescuers, but may simply be fellow employees trained to utilize certain specific equipment and procedures.

Selecting, equipping, and training co-worker assisted rescue responders is addressed in future chapters. This process can be a tricky proposition and should not be undertaken lightly. It takes a special person to help an injured co-worker, not just in terms of technical ability but also in terms of psychological resistance. Responders should be selected from those who are likely to be in close proximity

Figure 2.6 Co-worker assisted rescue may be as simple as moving a ladder into place.

to a potential subject, and should have not only the aptitude for the task, but also an appropriate attitude. These individuals should receive special training as well as frequent opportunity to practice their co-worker assisted rescue skills. Any time co-worker assisted rescue responders must assist a co-worker in an actual incident, follow up measures should be taken to identify and resolve the potential effects of post-traumatic stress disorders.

Co-worker assisted rescue may be as simple as moving a ladder or lift into place (Fig. 2.6), turning a crank, or pulling on a rope, without the responder ever exposing him- or herself to a fall hazard at all (i.e., non-entry rescue), or it may be more complex, with the co-worker assisted rescuer entering the vertical environment to remove the subject from potential harm, or even performing more advanced extrication and medical care. The vast majority of the time, co-worker assisted response will require intervention at a level somewhere between these two extremes.

The level of care rendered should be based on what the employer has prepared and authorized the co-worker assisted rescue responders to do. This means that responders should not attempt a rescue method for which they are not properly trained and qualified. Remember, the golden rule of rescue is to *not create further harm*, either to the subject or to rescuers. Responders should never render care beyond their level of training.

The level of training that co-worker assisted rescue responders should receive and the level to which they are authorized to render care is up to the "authority having jurisdiction," in this case the employer. These will be determined based on ability to develop and maintain skills, anticipated needs of the incident(s), and the amount of time it will take to connect a subject with more advanced care capabilities. Usually basic first aid and simple extrication, such as raising and/or lowering of a subject on his or her own system, will suffice. However, in the case where employees may get themselves into particularly complicated situations, or at remote work locations, the employer may choose to rely more heavily on the co-worker assisted rescue responders for more advanced skills. Employers who have more experienced personnel at their disposal, such as Society of Professional Rope Access Technicians (SPRAT) certified rope access technicians, will enjoy the benefits of more technical levels of access as well as rescue capability.

Provision for co-worker assisted rescue should be clearly outlined in the preplan, including information on equipment, anchorages, techniques, and expectations for extent of medical intervention. Co-worker assisted rescue is intended as an intermediary step; any time co-worker assisted rescue is called into action, the next level of response capability identified in the preplan should be notified and initiated. Rescue situations can change quickly or morph suddenly, and having the backup plan already moving into position will save valuable time when this occurs.

This next level of capability may exist in the form of onsite standby professional rescue, or offsite professional or municipal rescue resources. If offsite rescue can be *assured* to be capable of reaching the subject within the prescribed time limit (i.e., promptly), co-worker assisted rescue may be bypassed altogether. Notification of co-worker assisted rescue responders as well as professional rescue responders should be addressed in the preplan.

Professional Rescue

It is unacceptable to shirk adequate provision for personal escape and co-worker assisted rescue by relying solely on external municipal/professional rescue agencies to make up for the deficit. This is a grave mistake that has caused many a simple fall to become a more serious predicament due to delayed/inadequate response. In one extreme example of an incident where internal rescue was not in place, an unaccompanied worker was found dead, suspended from a platform by a safety belt.[20] His injuries included positional and compression asphyxia, abrasion and contusion of the lower abdomen from the belt compression, and hemorrhage in the neck and left clavicle area.

[20]Occupational Safety and Health Administration, Department of Labor, *Fatality and Catastrophe Investigation Summaries, Inspection: 123273278*, http://www.osha.gov/pls/imis/establishment.inspection_detail?id=123273278.

Professional rescue care is usually the dominion of privatized and/or municipal rescue agencies. At this level of response, rescuers are generally capable of rendering more advanced medical care (e.g., emergency medical technician [EMT] or paramedic level) and a broader spectrum of technical rescue capability. Some large employers, particularly those where a high likelihood of catastrophic incident is deemed likely (e.g., chemical plants) choose to employ internal professional firefighting and/or rescue services on a permanent basis. In such cases, employees are hired and assigned specifically and exclusively to the fire and/or rescue group. This level of professional rescue capability is relatively rare in the workplace.

Another type of onsite professional rescue capability might come from professional rescue services that are contracted to occasionally provide services on a temporary "standby" basis when high risk work is undertaken at a given worksite. An example of this might be at power plants, where personnel must periodically enter boilers or underground storage tanks for maintenance purposes.

The most common type of professional rescue comes from local municipal response agencies such as fire, search and rescue, and/or ambulance services. Although such agencies are generally trained to perform the types of rescues that are reasonably foreseeable in a given jurisdiction, what cannot necessarily be guaranteed is the municipal rescuers' familiarity with a specific worksite, their timeliness to respond, or their technical capacity to adequately resolve an incident in a particular environment. Not all rescue organizations are trained or experienced in technical, or high angle, rescue.

While local area professional rescue resources (Fig. 2.7) should be activated any time co-worker assisted rescue is called into action, employers should explore

Figure 2.7 Preplanning with local responders is essential. Credit: Pigeon Mountain Industries, Inc.

and clearly understand what this truly brings to the table, including anticipated response times (best/worst case), technical capabilities, and medical care.

Coordination with External Resources

A wise employer will choose to coordinate in advance with local emergency services provider(s). Most jurisdictional authorities appreciate having an opportunity to coordinate in advance with employers, particularly where unique hazards exist. Not all responding agencies are able to maintain proficiency in all skills, and if an offsite agency has not been made aware of specific risks posed at your worksite they may not be prepared to respond to an incident there with appropriate equipment and personnel.

Emergency services for your location may be provided by different agencies for different types of emergencies. For example, fire protection will usually be provided by the local fire department, but medical services may be outsourced to one or more private organizations. Law enforcement may fall under local or state police, or perhaps even sheriff's authority, depending on the type of incident and your location. Backcountry rescue, or rescue from remote worksites, often falls under the jurisdiction of law enforcement. You should contact your local emergency services department to determine which agency (agencies) are likely to respond to your facility for various types of incidents. From here you can contact each agency to determine what capabilities they do/do not possess.

During an actual emergency, the distinction between these organizations will most likely be transparent to you. As far as you can see, you just dial your local emergency number, describe the emergency, and in a little while someone shows up. However, a little advance planning can help ensure that those who do show up are actually prepared to deal with your particular emergency. It may come as a surprise that not all emergency responders are trained to render advanced medical care. You may also be surprised to learn that not all response organizations have the ability to perform rescue at height, or in confined spaces. There are many different kinds of emergencies to which a response organization must react and it is unrealistic for all agencies to maintain readiness for all types of incidents. Instead they must prioritize and choose those that are most critical and necessary for the area in which they work.

For this reason it is advisable to permit the agency (agencies) who are likely to respond to your location to review your fall hazard survey in advance so that they can evaluate their readiness, and be sure the person making the notification relays the nature of the event (e.g. high angle rescue, confined space, fire, emergency medical, etc.). Not all responding agencies have the resources to maintain an effective high angle rescue program, and even those that do may or may not be familiar with the equipment, techniques, and methods used for fall protection on your worksite. As an employer, you should review the hazards with the agency (agencies) to verify that they are capable of responding to all potential falls covered by the fall hazard survey, and/or use this information to make alternate plans if necessary.

Sometimes an organization will request advanced notification of certain types of hazardous work so that they can put special teams on standby. When in doubt, make that call.

If possible, it may be beneficial for the responding organization to perform training drills at your location. This offers massive benefit, and facilitates the most efficient exchange of critical information. Not only will this approach help to ensure that the agency is familiar with the fall protection equipment used on your worksite, since this equipment may differ greatly from the equipment commonly used in high angle rescue, it will also help you to know what equipment/techniques the agency will bring to your incident. Cooperative training days can also help mitigate management/authority clashes that might occur during a true emergency.

Another important thing to note is the maximum response time to areas covered by the managed fall protection plan. Note that there may be a difference between optimum response time and the amount of time it is likely to take if the agency is already busy with another call. Especially where a municipal organization is comprised of volunteers, and/or covers several types of emergencies (e.g., fires, auto accidents, and medical calls), response times can be compromised during busy periods. This may require response from another station or even a response by another agency, based on mutual aid agreements, and is something for which you should be prepared.

As an employer, your evaluation of any rescue service should consist of two elements: an initial evaluation to determine the adequacy of a resource for a given potential incident type, and a performance evaluation relative to an actual scenario(s) that might occur on the site. The latter evaluation should be performed at regular intervals to ensure continued capability. Readiness evaluation is outlined in greater detail in Chapter 10.

SUMMARY

Effective post-fall rescue capability is essential wherever authorized at-height workers are engaged. This capability is not something that creates itself. It requires intentional planning and preparation, requiring a combination of both technical ability (e.g., of ropes and rescue systems) and at least basic medical capabilities. While neither of these necessarily needs to be provided at advanced levels from internal resources, consideration should be given to both. A highly trained rigger or vertical technician is just that and should not be considered a "rescuer" unless he or she also possesses specific rescue and medical training to make him or her adept at patient evaluation, packaging, and care. Likewise, a paramedic or other medical professional is just that, and should not be considered a "rescuer" unless he or she also possesses specific rigging and rescue training to make him or her adept at extrication, evacuation, lowering, and raising techniques.

With a basic understanding of work at height and the fundamentals of developing a rescue preplan understood, Chapter 3 will delve into the program itself.

POST-CHAPTER QUIZ

Chapter 2

1. Wherever the likelihood of life threatening accidents is foreseeable, the employer is required to

a) Designate a rescue plan

b) Provide appropriate rescue equipment

c) Train designated employees accordingly

d) All of the above

2. The golden rule of rescue is

a) Let someone else do it

b) Do no harm

c) Do nothing until municipal responders to show up

d) Use a rope

3. True or False: OSHA will generally not cite an employer for failure to have a rescue plan in place.

4. True or False: Not establishing a preplan or assisted rescue plan is a good way for an employer to avoid being cited.

5. A written preplan for post-fall rescue

a) Is specifically required under OSHA regulations

b) Is the best way to ensure appropriate preparedness

c) Must be developed by an outside entity

d) All of the above

6. The three levels of response for a fallen worker at height include

a) Professional rescue, medical rescue, hospital

b) Personal escape, assisted rescue, professional rescue

c) Personal escape, employer rescue, fire department

d) Personal escape, professional rescue, ambulance

7. Identify the steps involved in developing a fall rescue response preplan (check all that apply).

a) Define area of responsibility

b) Identify known or foreseeable hazards

c) Develop specific rescue protocols

8. What should an EAP include? (Choose all that apply)

a) Escape procedures and personal escape route assignments

b) Rescue and medical duties for those employees who are to perform them

c) Means of reporting fires and other emergencies

d) Names or job titles of persons who can be contacted for further information or explanation of duties under the plan

9. What is one way to ensure that the appropriate notification takes place if a fall occurs?

____

10. True or False: Municipal rescue services always have the training, equipment, and ability to rescue anyone from any predicament that might occur in a workplace.

ANSWER KEY

1. d
2. b
3. False
4. False
5. b
6. b
7. a, b, c
8. a, b, c, d
9. Never work alone
10. False

CHAPTER 3

PROGRAM DEVELOPMENT AND INCIDENT MANAGEMENT

By the end of this chapter, you should be able to:

- Describe the function of a workplace response group.
- Understand why employers at a multi-employer worksite should cooperate in fall protection and post-fall rescue plans.
- Describe the order of priority for safety of the rescuer, co-rescuers, and the subject.
- Understand the importance of retraining and practice for co-worker assisted rescuers.
- Explain the difference between co-worker assisted rescue and standby professional rescue.
- Describe how to achieve essential components of incident management.
- Understand what kinds of knowledge and skills are crucial for responders to possess.

Assuming that you have developed, or are in the process of developing, a post-fall rescue plan as described in Chapter 2, this chapter will help you turn that plan into a co-worker assisted rescue program, and will also identify useful concepts in *workplace response group* management.

A workplace response group is made up of men and women who are designated to provide co-worker assisted rescue in the event of a fall emergency. The actual methods and techniques for co-worker assisted rescue are addressed in later chapters.

The personnel who comprise a workplace response group and the concepts and methods they use to perform co-worker assisted rescue are very different from their professional rescue counterparts. While professional rescuers receive extensive training and ongoing education in medical and technical aspects of rescue, the individuals in a workplace response group are typically performing this task outside the scope of their normal work responsibilities.

Depending on the level of capability that an employer feels it is appropriate to develop for a given location, development of a co-worker assisted response capability can serve as a complete solution, or simply to fill the gap between when an

Falls from Height: A Guide to Rescue Planning, First Edition. Loui McCurley.

incident occurs and the arrival of professional rescuers. Either way, much can be done in this time to contribute to a successful outcome for a person who has suffered a fall. Members of a workplace response group will be familiar with the hazards, work practices, equipment, and operations specific to a workplace, and as a result can often have a subject extricated from their predicament even before professional responders arrive. Professional rescuers, well-trained and skilled as they may be in their profession, cannot possibly be as familiar with a given worksite as the workers who experience it every day.

A commitment to developing co-worker assisted rescue capabilities through an internal workplace response group should not be undertaken lightly. Aside from initial training and outfitting, an employer must take into consideration ongoing requirements for keeping responders properly equipped, motivated, and proficient in necessary skills. In the previous chapters we discussed the importance of developing a rescue plan wherever a fall protection plan is used. As we progress toward developing specific personnel capabilities, it should also be made clear that in some cases an employer may be wise to include planning for a cooperative response where multiple-employer worksites exist. Excellent guidance in this area can be found in confined space regulation OSHA 29CFR1910.126, which also refers to working at heights.

Certainly it can be argued that an employer is ethically responsible for all workers on sites that they own or control. Under OSHA's "Multi-Employer Workplace Doctrine," while not actually specified by a particular rule, regulation, or standard, OSHA can, and sometimes does, cite general contractors for violations committed by subcontractors. This responsibility can extend to outside contractors coming to the employer's worksite to perform all manner of services, including electrical, plumbing, excavation, or structural repairs. That is to say, an employer can also potentially be legally responsible for the safety and health of a subcontractor's employee, if the employer creates the hazard, exposes the employee to the hazard, is responsible for correcting the hazard, or is the controlling employer on the site.

Although the obligation that a *controlling employer* has to prevent and detect violations for a subcontractor is different than what is required of an employer with respect to his or her own employees, still it is prudent for an employer to protect him- or herself as reasonable against potential liability. Here are a few things that an employer can do to help protect his or her interests when it comes to fall protection and post fall rescue:

1. Conduct a job hazard analysis of any owned and/or operated facilities to determine if and where fall hazards may exist.
2. Develop a standing access and fall protection plan for employees and outside contractors who may need to access to or work within the areas where hazards are identified.
3. Clearly define a process for what happens in the event of a fall, including all of the steps outlined in Chapter 2.
4. Conduct or arrange documented training for Authorized Persons who will access those areas. Arrange documented meetings with outside contractors (as

applicable) to ensure that they have been informed of the hazards and requirements, and that they are aware that they are responsible for ensuring compliance for their employees.

5. Monitor and enforce compliance, and discipline employees who are found to be in noncompliance.

Doing these things will decrease the risk of an accident, a citation, or a lawsuit while enhancing the safety culture—which in turn will reduce the controlling employer's potential liability exposure.

In addition to the legal responsibilities involved in prompt rescue after a fall, employers must be sensitive to the moral aspects of responsibility as well. If an incident occurs, the natural human response is for co-workers to want to help. Without adequate planning and advance preparation, this natural desire to come to the aid of another can be deadly.

In one such case, three out of seven co-workers were overcome, two of whom died, while investigating a problem with a backup in a sewer system grease interceptor/trap at a large hotel.[1] The crew determined that the blockage was located after the elbow in a section of PVC piping in a pit, so one employee entered the pit and began to cut through the PVC pipe with a reciprocating saw. No air sampling or ventilation was performed before or during the work. When liquid was released and began to flow out of the pipe, the employee tried to exit the hole but was overcome and fell to the bottom. Another employee entered the pit to save his co-worker. Although he was able to reach the first man and pulled him out of the water, he too was overcome and fell unconscious. At this point a third employee entered the pit but he, too, was overcome. Someone called 911 while others collected fans to ventilate the hole and searched for a rope. Emergency responders pulled the three men from the pit. Only the third man survived.

While this kind of incident is generally associated with the best of intentions, the inevitable adrenaline surge that accompanies a true emergency can overcome all logic and attention to safety. In a battle between logic and emotions, emotions will almost always win. A person will often to throw caution to the wind in order to provide immediate assistance to another unless he or she has been trained to be able to overcome that natural urge.

Any worksite where the expectations are unclear or poorly defined is a site where employees are unnecessarily exposed to risks associated with the inevitable spontaneous response that will likely take place in the event of an accident.

EMPLOYEES AS RESPONDERS

Where an employer has established a need for workers at height to use some means of fall protection, and where a potential fall (into a harness or onto ground level)

[1]Occupational Safety and Health Administration, Department of Labor, Fatality and Catastrophe Investigation Summaries, Inspection: 310429386. Accident: 201042942, Two Employees Killed, One Injured after Entering Manhole.

remains possible, three levels of capability should be researched and established as necessary:

1. Personal escape
2. Co-worker assisted rescue
3. Advanced/professional rescue

Most employers will identify personal escape and co-worker assisted rescue as reasonable in-house capabilities to develop and maintain. Personal escape capabilities are addressed in detail in Chapter 6. While these should be maintained through specific, ongoing training, success in co-worker assisted rescue efforts also requires attention to good teamwork. Teamwork in co-worker assisted rescue begins with careful selection of rescuers and then good management of the response group and its operations.

In any rescue situation, safety is paramount. When selecting personnel who will be assigned and trained in co-worker response, select those who are familiar with the workplace, who possess a thorough understanding of the hazards and associated protective measures, and who will maintain a cool head under extreme pressure. Responders should be trained to operate safely in accordance with established protocols and with an eye to whatever OSHA regulations may apply to their situation. If a rescuer allows him- or herself to become injured or compromised, it not only takes that responder out of the resource pool, it also compromises the entire operation as other rescuers attend to their fallen peer.

Each responder should be focused primarily on his or her own safety, and secondarily on that of the co-responders. The safety and imminent rescue of the subject on whose behalf the effort is undertaken, while important, ranks last in the priority queue. While rescue may require the rescuer to take certain calculated, controlled risks, every precaution must be exercised to maintain the safety of the rescuer. A "hero mentality" only creates additional problems and hazards for everyone involved.

With the fall protection plan in place, and employee roles thoroughly defined, the employer should ensure that employees and subcontractors are trained to whatever level of response they may be expected to undertake. Protocols and operational criteria appropriate to the training and experience levels of the organization should be developed and followed for the safety of all.

In the not-too-distant past, there was a surge among fire departments to create special rescue teams capable of high-level technical rescue response in various disciplines (rope rescue, confined space, trench and excavation, etc.) at several stations or locations within an area. While such teams are relatively easy to train initially and are even quite reasonable to equip, time has proven them unreasonably expensive and difficult to maintain in terms of training and operational experience unless they train and/or respond frequently. As a result, many professional agencies have subsequently decommissioned their special-response teams only a short while after standing them up. With a little more forethought and planning, the time, effort, and money that went into the standing-up of these resources could have been expended more productively.

This lesson is one that an employer can (and should) learn from. The lesson? Prepare for what you need. Not less, not more. And share resources whenever possible.

As a foundation, it is highly advisable for *all* workers to possess at least rudimentary knowledge of CPR and first aid so that they can assist their co-workers as necessary. No one wants to experience the agonizing situation of being at the site of an injury or illness and not knowing what to do. Some parts of regulation, such as the OSHA confined space regulation, require that at least one responder be certified in first aid and CPR. This text is not a medical reference, but numerous resources exist for basic first aid/CPR training, including resources from the American Red Cross, the American Heart Association, the National Safety Council, and the National Ski Patrol. The employer's written emergency action plan should provide more thorough guidance for recommended actions and responsibilities in the event that a worker falls.

Selecting Co-worker Assisted Rescue Responders

Post-fall rescue responders are best selected from staff who are already familiar with working at height. Working with a certain amount of air under one's feet while encumbered by personal protective equipment, and familiarity with equipment and systems used for rescue, is an acquired taste and can be disconcerting to the uninitiated. Selecting responders from people who already work at height can help mitigate this problem. This approach offers the added advantage of helping to ensure that the would-be rescuer is familiar with the conditions and equipment he or she is likely to encounter during the rescue.

In addition to aptitude for the job, there is a certain attitude that is conducive to a good rescuer. Fearless, adrenaline-pumped, wannabe superheroes are seldom a good choice for such roles. Rescuers must be good team players (especially proficient communicators), confident but not cocky, and able to assess risks and solve problems safely on the fly. In short, an individual assigned to the role of "rescuer" must inherently possess a certain level of maturity in order to maintain composure under stress, make difficult decisions with wisdom, and manage the responsibility of life-saving duties.

In some instances, rescuing a fellow worker can be physically demanding. The level of physical fitness required for the job will vary depending on the specifics of the worksite and potential hazards. Ideally, these individuals should be in general good health and free from serious physical deficiencies. This may include establishing basic medical criteria as well as requirements for the ability to perform certain physical skills that are consistent with the type of activities that might be required in the event of an incident.

While there are typically not specific educational requirements that co-worker assisted rescuers must meet, they should be capable of reading and comprehending product instructions, procedural manuals, and other materials related to safety and rescue equipment, methods, and training. They must also be capable of communicating effectively and cooperating well with teammates.

Forming Co-worker Assisted Rescue Capabilities

Formation of a co-worker assisted rescue program should focus on the safety of the worker. This may seem to be a self-evident statement, but it is worth emphasizing that the true goal in developing a post fall rescue program is that of safety, not just an exercise in "checking a box." Everyone involved should recognize that these efforts represent proactive measures that are being taken for the protection of everyone who is working at height.

The idea is to ensure that *any* worker who falls can be rescued. The time and cost of providing each person working at height with a personal escape capability as well as a basic co-worker assisted rescue capability is minimal by comparison to the potential cost of an incident where prompt rescue is not performed. Such training will help increase employees' fall protection awareness and will help ensure emergency preparedness in your facility.

This text was developed expressly for the purpose of assisting employers in developing personal escape and co-worker assisted rescue capabilities, which *every* at-height worker should have. It is simply not acceptable for just one or two individuals per shift to receive such training, especially if that person is also working at height. What if *that* person falls: who will rescue him or her? *Every* individual working where fall protection/fall arrest equipment is used should be trained in what *not* to do, as well as what to do, in case his or her buddy might fall—even if the planned methods are as simple as moving a tall step ladder into place to allow the fallen worker to recover him- or herself.

For a slightly greater response capability without committing to a full-blown professional rescue operation, a few volunteers can often be solicited from personnel who are already working at height to receive additional training to serve on an internal workplace response group. Personnel who are willing to commit a few extra hours per month to rescue preparedness will usually be relatively easy to find in these situations, and will often step forward on request. Providing them with quarterly training, as well as support and recognition, will help improve safety in your facility, and will give these workers and others a feeling of increased confidence and security in their daily work.

Where the probability and/or consequence of a rescue are extremely high, or where multiple complex hazards overlap (e.g., fall protection, confined space, hazardous materials, contaminated air), immediate response at a higher level of professional rescue skill may be required. In these cases, utilization of an on-site *professional standby rescue team* may be warranted. Professional standby rescue can be outsourced, or can be developed internally. The capabilities of a professional standby rescue team are usually selected and tailored specifically to the needs of the worksite, and include both advanced technical rescue capabilities as well as medical capabilities to at least the emergency medical technician (EMT) level. This is a much more significant undertaking than co-worker assisted rescue, so temporarily contracting rescue services to an external firm who specializes in this type of work is often the best choice for an employer who has short-term needs.

Needs Assessment

As a first step, provision for accessing a fallen worker within a reasonable time frame must be made. The amount of time deemed "reasonable" to complete the rescue will depend on circumstances and is based on the requirement to remove the patient from the predicament before further harm. As discussed in previous chapters, 15 minutes can be considered a reasonable maximum target response time. Whether it is sufficient to merely make physical contact with the subject within 15 minutes, or if the entire rescue needs to be effected and completed in that time, will be dictated at least to some degree by the degree of hazard posed. Keep in mind, however, that studies indicate that motionless suspension in a harness can be detrimental within that amount of time.[2]

It bears repeating that this 15-minute response capability will very rarely be achievable solely by utilization of local municipal rescue resources. Emergency services throughout the world are provided through a combination of professional and volunteer resources. Most of the geographical United States is served by volunteer services, while the higher population areas are usually served by professional rescue. Many professional departments strive for a 5–10 minute response time to the addresses within their respective jurisdictions, but this goal typically pertains to the arrival of fire suppression apparatus at the curb outside, not necessarily to placing rescue-capable response within the facility. In short, normal municipal response times are simply not fast enough for most fall protection rescue needs—especially if the subject is not at the front door of the workplace!

Employers should also understand that a given municipal response agency may or may not be willing and able to respond to every possible type of incident in its area. As with all employers, the first priority of the rescue agency is to protect its own employees. This includes clearly defining its own response capabilities and working within those parameters. Emergency response organizations are as responsible as any other employer for prioritizing appropriate safety and care of their workers. In the interest of safety for all, employers should coordinate in advance with local emergency services providers to determine whether responders are prepared to deal with hazards found in their workplace, or whether such hazards might be considered extraordinary.

Assessing just how much assisted rescue capability for falls is needed in your facility begins with a review of the fall protection and rescue plan described in Chapters 1 and 2 of this manual. It is reasonable to suggest that any organization employing personnel to work at height ensures that those employees, and those working near them, should be capable of at least recognizing when an incident has occurred, knowing how to activate appropriate resources, and being able to stay out of trouble themselves. These things are not necessarily as easy as they sound. Injury and death in the line of duty is all too often associated with the initial reaction of a would-be rescuer being inconsistent with his level of preparedness. This is especially prevalent in confined spaces, where more than 60% of fatalities occur among would-be rescuers.[3]

[2]U.S. Department of Labor Occupational Safety and Health Administration Safety and Health Bulletin Suspension Trauma/Orthostatic Intolerance (March 24, 2004; updated 2011), http://www.osha.gov/dts/shib/shib032404.html.

[3]DHHS (NIOSH) Publication Number 86-110, http://www.cdc.gov/niosh/docs/86-110/.

If a local rescue agency cannot ensure a timely response/rescue from each location where a fallen worker might be, even when they are on another call, and/or if the local agency is not equipped or prepared to ensure rescue of a fallen worker from any foreseeable situation in a given workplace, co-worker assisted response is more than just a good idea; it is a duty.

How far the capabilities of such a response should go is dependent on how quickly a fallen worker needs to be rescued to avoid further complication, how potentially complex the extrication might be, whether or not nearby municipal/professional rescue is available and able to assist, how long it might take for professional rescue to arrive, and how feasible it is for the employer to maintain a given level of co-worker assisted rescue capability once it is established.

At the very least, co-worker assisted rescue responders should be capable of sizing up the rescue scene, setting into motion appropriate actions to remedy the situation through personal escape and/or co-worker assisted rescue, identifying and activating the next level of response, establishing a management structure for the incident, mitigating hazards on an immediate level, and ensuring the use of appropriate personal protective equipment (PPE). These things seem few on the surface, but maintaining sufficient training in even these categories can command significant amount of training time.

Organization and Group Development

Once a decision is made to maintain some level of response capability, someone must take responsibility for ensuring that there are field-able personnel on site and available whenever a worker is at risk. Normally this responsibility falls to the site safety manager, but ultimately it is up to the employer or his or her designee.

Co-worker assisted rescue responders should be capable of utilizing equipment and systems defined by the employer, and of removing a fallen co-worker from further danger in accordance with pre-established methods. This calls for some representation of a management structure, in which an individual or individuals become responsible for the operations, training, and upkeep of the internal response group.

Personnel assigned to perform co-worker assisted rescue responsibilities should be provided adequate guidance on expected outcomes and procedures, proper equipment, appropriate training, and sufficient time to practice their skills on a regular basis. The advice of a resource who is experienced in rescue is key at this phase. A voice of experience will also be useful when it comes to establishing protocols and training criteria that are appropriate to the defined capabilities and experience levels of the organization. There is no single set of skills or systems that can be effectively applied to all workplaces. Each workplace must be independently evaluated to determine specific hazards, personal skills necessary, and appropriate response protocols. While there are some generalities that can be addressed insofar as equipment and systems are concerned, the most appropriate means for a given rescue are quite subjective and dependent upon environmental conditions, nature of the incident, height considerations, rescuers available, and other factors.

An organization with a co-worker assisted rescue plan should appoint a leader who has enough depth of experience and training to make good decisions in this regard, and ensure that this leader has access to appropriate experts and resources in the field.

The appointed rescue leader should ensure that all supervisory functions are fulfilled with regard to staffing, training, and executing rescue preparedness. This is not to say that the rescue leader must perform all of these duties, only see to it that they are performed.

Operational Organization

Workers who are responding to a co-worker in need are not likely to be professional rescuers, and the organization in which they function is not likely to be a professional rescue organization. Nonetheless, it is important for employers to develop a functional organizational management system that will be effective in the event of an emergency, and to recognize that this management system may differ from how the organization is managed on a day-to-day level.

The chain of command during emergency operations in most cases should differ from the chain of command during day-to-day operation. This is because management of rescue operations requires special knowledge and training. The many layers of management often found in day-to-day business operations are not generally conducive to a rescue operation, and in fact can render a rescue operation ineffectual. It is dangerous to try to run a rescue by committee, or for a rescue to be run by a supervisor who is not specifically trained in rescue.

Those who may be accustomed to "being in charge" of business operations should be clearly informed and understand *in advance* as to the nature, extent, and limitations of their responsibilities during an incident. Briefing executives and managers before an incident ever occurs, and having them sign off on the emergency operations guidelines, will help prevent misunderstandings and power struggles during an incident if an emergency situation arises.

In the world of professional rescue, incident management is generally achieved using some form of what is known as the *Incident Command System (ICS)*. The prevalent use of this system is both a testament to its success, and justification to emulate it even in the co-worker assisted rescue phase of a paraprofessional workplace response. If effective incident management procedures are implemented early, from the first efforts of a co-worker assisted response, it can adapt easily if the incident should escalate to a more complex professional rescue response—which should be initiated immediately whenever in house efforts are begun. A well-thought-out system will expand logically as the need grows and should be able to be utilized for any type or size incident ranging from a minor accident with a few employee responders to a major situation involving several municipal or professional agencies. No matter how small or how big a situation may be, the ICS model can be used effectively as an incident management tool.

There are several considerations that should be taken into account when developing an internal incident management system. Years of experience in emergency management shows that the following criteria are all essential concepts in managing emergencies:

1. Common Terminology

   In order to reduce confusion, a pre-established command system will use common terminology for organizational functions, resources, and facilities. Consider consulting with your local emergency services provider to determine what terminology they use, and consider mimicking it where appropriate.

2. Modular Organization

   The organizational structure of an incident management system should be set up to grow in a modular fashion based upon the kind and size of an incident. Key roles and responsibilities include:

   (a) Incident Oversight: First and foremost, someone should be designated as "where the buck stops", right from the initial phases of any response. The professional Incident Command System calls the person who performs this function an *incident commander*. You may call this person whatever you wish so long as someone is pre-established and identified as "in charge"; it may be the president or CEO of the company, or it may be someone with more specific rescue management training or experience. This is the most important position in the structural organization, but the person doing it should not be the one of the people who is engaged in the operational phase of the rescue. This person should remain free of operational tasks and should have full company authority to make big-picture decisions. In addition to being responsible for the effective overall management of the incident, key responsibilities of the incident commander include safety, information, and liaison with other resources. The incident commander may wish to solicit assistance from others in the following forms:

      - Safety Officer: If resources permit, assigning a safety officer (or "safety attendant") who is experienced, knowledgeable, and reports only to the incident commander can help assure personnel health and safety. There is a lot to consider during an incident, and having someone assigned to assess and/or anticipate hazardous and unsafe situations on an ongoing basis can help avoid something being overlooked. The safety officer should work with the workplace response group to develop a site safety plan in advance, and to help keep it updated it during an incident. This individual will review and monitor the rescue plan on an ongoing basis for safety implications, and provide ongoing assessment of hazards and required controls.
      - Information Manager: Information management is crucial, and utilization of trained public relations professionals in this area can help protect the organization during the incident and potentially long afterward. An individual (or individuals) should be assigned to collect, create, and disseminate information about the incident outside the organization. This staff should be closely integrated into the management structure so

that they can effectively provide ongoing news (consider using press releases), updates on activities carried out by responders, and communications regarding issues that may affect the community. Crisis media management is a general failing for many industrial sites. A common response to fear is often to refer all media inquiries to specialists who excel in stonewalling and non-dissemination of facts. This approach is counterproductive and ultimately serves only to fuel the fires of misinformation; instead consider allowing those who serve in public relations roles to work closely with corporate executives and/or legal advisors to deliver timely, accurate facts.

- Agency Liaison: Assigning an individual who is familiar with and can serve as the point of contact for external agencies and groups can help keep the incident commander's distractions to a minimum. This individual should be briefed in advance and should be adept in communicating with political figures, local government officials, and even criminal investigators who might begin arriving on the scene even during the incident.

(b) Incident Operations: This individual is responsible for direction and coordination of the tactical, hands-on functions of the response in accordance with the overall wishes of the overall incident manager. The operations manager should be in direct communications with the incident commander and should be able to translate their objectives into clearly defined and assigned tasks. They should be well versed in available equipment, systems, techniques, and personnel so that they can make appropriate decisions and assignments.

(c) Administrative/Legal/Financial: The administrative section is important for monitoring of the administrative aspects of the incident and of decisions made relative to the incident. This includes the financial, regulatory, and legal aspects of the operations, and may involve decisions that are directly related to the rescue operations, as well as loosely related matters (such as whether to stop work in the facility or send other workers home). A high ranking company executive should manage this area, and the assistance of a legal advisor (before and/or during an incident) is invaluable.

(d) Planning: The person(s) responsible for planning keeps tabs on how things are going with the incident, and then projects and forecasts next steps and future resource needs on an ongoing basis. This is not the preplan (which hopefully already exists), but the ongoing operational plan. This includes tracking the status of every piece of equipment on the incident, predicting future needs, maintaining access to external resources, updating public relations staff, and cleaning up after the incident. Documentation of things as they happen is an important part of incident planning.

(e) Logistics: The logistics section manager is responsible for ensuring that the practical needs of the operation are fulfilled and functional. This includes such things as communications systems, transportation (if needed), services, personnel, equipment, materials, and facilities, as needed.

Because the incident commander is responsible for everything during an incident, it may be helpful for him to also appoint assistants to manage one or more of these functional areas of the response. For discussion purposes we will refer to these as functional sections, which is consistent with professional ICS terminology. The incident commander may delegate one or more of these, or may handle any number of these roles him- or herself.

While a general industry or construction employer may use different terminology from that described above, in effect all of these responsibilities must be managed during an operation. In essence, operational management could translate to a construction or industrial site as follows:

Incident oversight/command: Company owner, CEO, or other executive

Administrative/legal/financial: VP or other executive, and/or legal advisor

Planning: Manager-level staff with rescue experience

Operations: Rescue group leader

Logistics: Manager level staff or administrative assistant

Public relations: VP of communications

Safety officer: Plant safety manager

Agency liason: VP

3. Integrated Communications

Communications are best managed through the use of a prearranged communications plan that everyone understands. If necessary, two-way radios may be used but be sure to provide training to anyone who may be using radios. While this seems like a simple matter, effective use of a radio requires more than just picking it up and talking. Potential users should be trained in, and practice, the use of crisp, clear language, radio etiquette, and proper operation of radios before having to use them in an emergency. Any channels used for an emergency should be dedicated as such, and should be cleared at the outset of an operation. A single point of contact, usually in close proximity to the individual who has overall responsibility for the incident, should be identified as the communications center/authority. Communications should be confined only to essential messages, and should be in plain English. No codes should be used.

4. Unified Command Structure

When rescue resources are expanded by the arrival of community response agencies, such as the arrival of ambulance, fire, or rescue, these will need to cooperate closely with the workplace response group and with one another. Integrating key personnel into the command structure will help ensure overall incident objectives are established and met, appropriate strategies are identified and used, joint planning for tactical activities is achieved and maximum use is made of all the resources at hand. Most municipal response organizations utilize terms such as incident commander, section, group, and team, which should translate reasonably well to the guidance offered in this text.

5. Consolidated Action Plans

Every incident needs some form of an action plan for successfully resolving the emergency. In the industrial environment, this plan will ideally consist of a written post-fall rescue preplan. During an actual incident, this plan becomes a living document and is focused, modified, and expanded as necessary, in accordance with circumstances and resources. Sometimes, where a preplan is not available, the entire plan may be developed in real time. In other cases, although this is not a recommended approach, the plan may never even find its way into writing. Whether a plan is written or not, care should be taken to ensure that the plan and any updates are effectively communicated among all who need to know.

6. Manageable Span of Control

The term *span of control* refers to the number of resources that can be adequately managed by one individual. In general, it is recommended that the span of control of any individual with emergency management responsibility should range from three to seven—with five being considered optimum. An important consideration in span of control is to anticipate change and prepare for it. This is especially true during rapid build up of an incident when good management is made difficult by the addition of resources. So, while an internal response group may consist of a larger number of people, say 6–12 individuals, during rescue operations it may be advisable to break this group into two or more logical teams, each assigned to a given task and reporting to a single team leader, who in turn reports to the incident commander. For example, one team may be assigned to rig the rope system while another team is assigned to access the subject or provide patient care.

7. Predesignated Incident Facilities

Even something as simple as knowing where an operation will be managed from can help not only responders but also bystanders, external resources, public relations personnel, and others. Whether people need to simply stay out of the way, or if they need to know where to go to report in, having one location serve as the central operations center will help focus the operation. For the industrial environment, it is best to designate this location in advance, in the emergency action plan. Be sure to include provision for expansion of the incident, such as where municipal emergency services should park vehicles and stage when they arrive, and where public information and news releases will be made available.

8. Comprehensive Resource Management

A *resource* is considered either an individual or a group of individuals who have a specific purpose within the context of a response. For example, your workplace response group as a whole can be considered one resource. When ambulance personnel arrive, together they too may be considered a resource.

To make most effective use of resources, the management group must maintain an up-to-date and accurate picture of what resources are available and which are deployed at any given time. All resources should be assigned a current status condition, and all changes in resource locations and status conditions should be made known promptly to the appropriate section manager.

ESSENTIAL COMPONENTS OF INCIDENT MANAGEMENT

1. Common Terminology
2. Modular Organization
3. Integrated Communications
4. Unified Command Structure
5. Consolidated Action Plans
6. Manageable Span of Control
7. Predesignated Incident Facilities
8. Comprehensive Resource Management

With careful selection and preparation, an in-house response capability can be as effective as a well-oiled machine and can ultimately mean the difference between life and death for a fallen worker.

Preparing Responders

Once a preplan is firmly in place and response protocols are established, rescue responders should be trained according to that preplan. Developing true teamwork within the context of co-workers engaged in potential rescue involves a combination of appropriate equipment for the intended task and proper training of personnel in skills and systems.

Equipment, skills, and systems will be addressed in later chapters. The aspect of proper training will involve ensuring that all personnel know, understand, and can work within established protocols to achieve the desired level of performance. Training is an ongoing need, and involves more than just an initial introduction. Initial training generally takes an initial commitment of at least 40 hours for each individual who has a role, and at least another 8–16 hours quarterly to maintain those skills. Training and re-training should include a review of relevant concepts, skills reviews, practical sessions including *mock scenarios*, and an evaluation.

It is good practice to document not only the training itself, but who attends each training and the results of skills evaluations for each person. The importance of re-training cannot be over emphasized. Research cited by National Training Laboratories in Bethel, Maine, suggests that as much of 70% of a person's knowledge is lost within 2–3 weeks of acquiring that knowledge if it is not used.[4] Put another way, the average level of retention for a given topic is approximately 30% as little as one month after training!

The above cited data do not take into consideration the further effects of adrenaline. The "fight or flight" response induced by the sympathetic nervous system during a high stress situation can sometimes have rather unpredictable effects on a person's actions, ranging from causing a person to freeze up to inciting irrational, extreme behavior. Performing drills to help responders learn whether their own natural tendency is to fight or to take flight, and to help them learn to cope with stressors, can help them learn to respond more effectively under such stressful circumstances.

Training more often for shorter periods of time is one way to help ensure maximum success with minimal investment in training and retraining. Be sure your training includes all aspects of rescue response, including management protocols, technical skills, medical considerations, and psychological preparation.

Contracting with an outside rescue training organization to help you develop your plan, determine your response protocols, and maintain personnel skills is an excellent way to maintain perspective and avoid a myopic view. Don't change external resources too often, however. There are many "correct" ways to do most things, and different experts will tend to have different preferences and styles. Consulting with different trainers or consultants just for variety's sake can backfire by leading to inconsistency and potential confusion. Find an expert you trust and stick with him or her, varying only to confirm and verify practices. One good method of doing this is to have your entire program—both preplan and field practices—audited periodically by an independent expert, allowing him or her to provide critical review and recommendations for improvement.

Knowledge and Skills

The co-worker assisted rescue responder should be trained in some basic concepts about responding to an incident before delving into the more tactical/performance aspects of the job.

Size-up *Size up* of a rescue incident is one foundational skill that any co-worker responder should possess. During size up, the responder should be able to:

1. Determine whether the scene is safe.
2. Determine the scope of the incident, including
    a. Physical area
    b. Number of potential subjects
    c. Number and location of bystanders
3. Note the last known location of the subject.
4. Evaluate the surroundings.

[4]E. Dale, *Audiovisual Methods in Teaching*, New York: Dryden Press, 1969.

5. Make a quick determination of what type of protective equipment is needed.
6. Form a quick analysis of progressive hazards (weather, environment, time of day, medical condition of the subject).

Co-worker assisted responders should be trained to manage initial incident hazards so as to minimize risk to rescuers and victims as a part of the sizing-up process. Anyone who might have witnessed the event should be identified, interviewed, and isolated in case further questions become necessary. Bystanders should be kept away from the hazard zone, and responders should use the buddy system to ensure that co-worker responders utilize adequate personal protective equipment.

Initial Planning Formation of an initial plan is the next step that a co-worker assisted responder should be able to take. The plan will normally be based on the individual's knowledge of the preplan, and may include such steps as:

1. Activating emergency response protocols.
2. Making a quick determination of what rescue methods will be used.
3. Assessing availability of personnel resources and their capabilities.
4. Assessing equipment availability.
5. Estimating the time to arrival of the next level of response.

In assessing what type of rescue will be needed, responders must understand the type and nature of incidents that might occur in their workplace and what the established preplan calls for as standard operating procedure, as well as limitations of their own response capabilities. The ability to make a good risk-benefit assessment based on the viability/survivability of the subject is also useful.

Resource Management Managing resources for effective operations requires training in practices and procedures to ensure:

1. Task assignments are communicated and monitored
2. Standard operating procedures are followed
3. All personnel are accounted for
4. Incident actions are documented
5. Command structure is followed
6. Rescue efforts are coordinated, and
7. All efforts are consistent with applicable regulations.

Special training in incident management may be useful for those who may be called upon to perform these roles. Resource management can become quite a challenge if the incident becomes large or complex, or if the co-worker assisted responders are likely to need to integrate with professional responders when they arrive. A number of forms and documentation tools are also available to assist with resource management, including those found in ASTM F1767.[5]

[5]ASTM International, 100 Barr Harbor Drive, P.O. Box C700, West Conshohocken, PA, 19428-2959.

Search Capabilities An element of search may be required if the subject is not readily visible or if there is question as to their precise location. Conducting a search requires coordination, special skills, and planning. It also requires the ability of an incident commander to effectively utilize personnel who are trained in investigative techniques and clue awareness. When conducting or performing a search, it may be important to know how to:

1. Establish and communicate search parameters.
2. Establish, communicate, and understand a victim profile.
3. Manage the entry and exit of all people who may be within the search area.
4. Effectively question anyone who may have viable information, and use that to update the knowledge base.
5. Assign/perform assignments properly and expeditiously.
6. Minimize risks to searchers.
7. Constantly maintain awareness of location of all searchers.

Anyone assigned a search task should know how to safely enter, maneuver in, and exit the search environment without creating additional hazard to him- or herself or others, and should at all times be capable of performing personal escape if necessary. Personal protective equipment should always be used, as required, and results of the search effort should be reported to whomever is in charge of managing the incident.

Machinery and Equipment Safety Familiarity with equipment and machinery used in the workplace is essential before attempting any rescue in that place. This will help facilitate the rescue, and may help prevent the responder from becoming a victim. A workplace responder should, at a minimum, be able to:

1. Identify and evaluate hazards associated with machinery within the workplace.
2. Comply with operational protocols for equipment to which one may be exposed.
3. Determine the required fire suppression and safety measures associated with machinery and equipment.
4. Select and use appropriate personal protective equipment.
5. Operate emergency shutoff control devices.
6. Know where chemical, superheated or -cooled lines, electrical, radioactive, or fire and explosion hazards exist on the site.
7. Be familiar with onsite fire extinguishment or control options.
8. Apply and operate stabilization devices used at the worksite.
9. Identify access/egress points from machinery and at site.

Subject Care Care and transport of a subject is ultimately the purpose of rescue. While this text refers to the rescuee as a subject rather than a patient, and emphasizes rigging techniques and evacuation methods, this should in no way detract from the importance of caring for the subject as a nurse might care for a patient. The subject of the rescue is a human, and care must be taken to manage them with empathy and compassion. To that end, responders should be capable of:

1. Applying universal precautions for infectious disease.
2. Rendering the ABC's of basic first aid.
3. Performing layperson CPR.
4. Packaging a subject in a stretcher or patient transport device, if one is identified in the preplan.
5. Determining an appropriate egress route.
6. Safely moving a subject a short distance to remove him or her from danger.
7. If the employer's emergency plan calls for use of a transport device (litter), responders should know how to use the device on a flat surface and up/down low angle slopes or stairs where ropes are not required.

Termination and Debrief An incident is not completed until all personnel and equipment are accounted for and returned to a ready state. This should be accomplished as expeditiously as possible after an incident. Employers may have a number of tools to help with this, including checklists and forms, to ensure that the following steps occur:

1. Retention of equipment, documentation, or other legal requirements.
2. Damaged equipment is removed from service.
3. Used equipment is cleaned, replaced, repacked, and put into service as applicable.
4. Personnel are debriefed as necessary.
5. Opportunity for *critical incident stress debriefing* is provided to anyone involved.

SUMMARY

Co-worker assisted rescue capability can be effectively and economically achieved for virtually any workplace. Interested and capable individuals may be assigned to serve on the workplace response group, and will often readily volunteer. Especially where people work in a high angle environment, your workers care about their own safety and that of their co-workers. Helping them to help themselves helps everyone involved, and makes for a more productive, upbeat, and positive work environment.

POST-CHAPTER QUIZ

Chapter 3

1. True or False: In the case of a multi-employer worksite, only the subcontractor has responsibility for his or her direct reports.
2. Whose safety should the co-worker assisted responder consider to be his or her highest priority during a response?
   - **a)** The subject
   - **b)** His or her co-responders
   - **c)** Him- or herself
   - **d)** The CEO of the company
3. True or False: Only those who are listed on the co-worker response group roster should be trained in CPR.
4. Employers should consider creating a preplan for personal escape, co-worker assisted rescue, and advanced/professional rescue
   - **a)** Only if the local fire department is more than 10 minutes away
   - **b)** Only if an employee might be exposed to a potential fall
   - **c)** Only if administrative controls are used for fall protection
   - **d)** Only if an OSHA inspector tells them to
5. A needs assessment should begin with the employer determining the ability of resources to arrive at a fallen subject's location
   - **a)** Within a reasonable time
   - **b)** With appropriate equipment to perform a rescue
   - **c)** With appropriate personnel to effect a rescue
   - **d)** All of the above
   - **e)** None of the above
6. According to industry best practice for incident management, the essential components of incident management include:
   - **a)** Common terminology
   - **b)** The local municipal fire service
   - **c)** Consolidated action plans
   - **d)** Manageable span of control
   - **e)** a, c, and d
   - **f)** a, b, and c
7. Which of the following administrative roles are important to consider as part of managing an incident? (Choose all that apply.)
   - **a)** Incident oversight
   - **b)** Administrative/legal/financial
   - **c)** Planning
   - **d)** Operations

e) Logistics

f) Public relations

8. According to the research cited in this chapter, the average level of retention for a given topic is approximately _____ % as little as 1 month after training.

a) 90

b) 30

c) 70

d) 10

9. True or False: Size up of an incident includes assessing the number of people potentially involved.

10. True or False: The co-worker assisted rescuer should never perform life saving medical care.

ANSWER KEY

1. False
2. c
3. False
4. b
5. e
6. e
7. a, b, c, d, e, f
8. b
9. True
10. False

PART II

EQUIPMENT, SYSTEMS, AND SKILLS

CHAPTER 4

EQUIPMENT FOR CO-WORKER RESPONDERS

When you are finished with this chapter, you should be able to:

- Know what items comprise PPE.
- Explain (conceptually) the difference between PPE for rescue as compared with PPE for fall protection.
- Describe key requirements for PPE, including:
  - Harnesses
  - Helmets
  - Connectors
  - Lanyards
  - Rope grabs
  - Personal descenders
  - Personal ascenders
- Explain the relevance of equipment standards applicable to industry versus those applicable to professional rescue.
- Discuss what "compatibility" means, and why it is important.

Up to now, we have focused on the administrative and conceptual facets of rescue. This chapter begins to introduce some of the practical, hands-on aspects that will turn those concepts into reality. Here, equipment is explored both from a regulatory and a performance perspective. It is most applicable to those who will be using the equipment, that is, field personnel. However, those who may be tasked with authorizing and/or paying for equipment may also find the information in this chapter to be of benefit.

Equipment used in fall protection and rescue can generally be separated into two categories: *personal protective equipment (PPE)* and *rescue equipment.*

PPE refers to items typically *worn by a worker* to provide protection against recognized hazards, while rescue equipment consists of gear used to perform a rescue task. Depending on the job task to be performed, PPE might include such items as safety glasses, helmets, gloves, safety harnesses, and lanyards, while rescue

Falls from Height: A Guide to Rescue Planning, First Edition. Loui McCurley.

equipment might include such things as block and tackles, connectors, and rescue systems.

Anyone working at height in compliance with regulatory requirements will have a basic complement of PPE already. Although all of this equipment should meet appropriate regulations and requirements for safety in the jurisdiction where work is taking place, just because equipment meets safety requirements for fall protection does not necessarily mean that it is adequate for rescue purposes.

REGULATORY REQUIREMENTS FOR PPE

Items worn by an individual, or used directly to provide personal safety, may be referred to as PPE. A harness, for example, would be PPE, whereas a rope would not. In the European Union, PPE is categorically controlled under what is called the "PPE Directive" of the European Committee for Standardization (CEN).[1] Under this requirement, any item of equipment that falls into the defined category of "PPE" must meet certain basic criteria, including compliance with an appropriate standard. What constitutes whether a standard is "appropriate" depends on the type of work being performed. European standards, at least at the writing of this text, do not take into consideration different work applications but instead address all types of a given piece of equipment under a single "norm," or standard. Harnesses used for fall protection, for example, will meet EN 891, the same standard as harnesses used for rope access, or for rescue.

In Chapter 1 you read that this differs from the United States, where PPE that is used for different purposes is regulated specifically according to its use. The regulatory authority in the United States is called the Occupational Safety and Health Administration, or OSHA. OSHA sets forth rules and standards that employers must meet in order to protect the worker. You also learned in Chapter 1 that specific product testing and performance requirements are largely absent from OSHA documents.

Product testing is instead deferred to national consensus standards. These national consensus standards are generally promulgated by specific industry organizations and are relative to that industry. As a result, there may be several different standards that could potentially apply to one type of product. Using the case of the harness described above, in the United States there are at this writing at least three standards that may apply: ANSI Z359 for fall protection; NFPA 1983 for rescue; and ASTM F1772, also for rescue.

Requirements for similar products may differ between standards for different industries, such as fall arrest, rescue, rope access, and rope course work, because different types of work involve different hazards, different work methods, and different priorities. For example, it is possible to use a harness with only a dorsal attachment for some types of fall arrest, while a front attachment point (either sternal or waist) is a necessity for rescue.

For a generic standard to mandate very specific requirements that fit one application very well might in fact reduce safety if used in another. For example, steel-toed boots may be mandated for some workplaces, but would be dangerous for a worker who may come into contact with live circuits or electrical hazards. As you

[1]Comité Europeen de Normalisation, CEN-CENELEC Management Centre, Avenue Marnix 17, B-1000 Brussels; www.cen.eu.

can see, what might be deemed "safe" is largely dependent upon where, how, and in what capacity something (or someone) is employed.

In an attempt to provide guidance in criteria for equipment selection, certain professional organizations bring subject matter experts together to write standard test methods, standard specifications, and standard guides for equipment and processes. Today in the United States it is not uncommon for standards to be developed by organizations that represent the interests of specific industries, including the National Fire Protection Association (NFPA), the Mountain Rescue Association (MRA), Society of Professional Rope Access Technicians (SPRAT), American Society of Safety Engineers (ASSE), and more. Such guidelines are known as *industry consensus standards*. While often criticized for lacking external perspective, industry consensus standards frequently represent best practice from within a given field.

Some organizations in the United States choose to have some or all of their standards endorsed by the American National Standards Institute (ANSI), which acts as a third-party to oversee adherence to processes and procedures for a fair and balanced document. These standards are called *full consensus standards*. Although full consensus standards are still usually written from within a specific interest group, submitting the document for further ANSI procedural endorsement offers at least some assurance that steps have been taken to ensure a fair and reasonable representation from multiple types of inputs (users, manufacturers, test houses, etc.), opinion and viewpoints from outside that industry have been considered, and careful analysis has been made to ensure that every negative vote is given appropriate consideration. This process may or may not be any better than with an unendorsed industry consensus standard, but the difference is that a non-partisan entity (ANSI) has at least verified procedural conformity.

There is another type of standard called *general consensus standards*. The best-known organization promulgating this type of standard in the United States is ASTM International. General consensus standards have as their greatest weakness the very thing that is their greatest strength: anyone and everyone is invited to participate in the development of any standard. While at first blush it may seem ridiculous that a veterinarian may be involved in writing standards about fall protection equipment, the truth is that the veterinarian may offer a different perspective that does indeed have some value to add to the fall protection standard, and more participation in the process generally adds up to greater quality. ASTM International also controls the balance and process of these documents, just as ANSI does, ensuring that proper processes are followed, negatives are addressed, and consensus is reached. Because of the wide representation and review, ASTM standards can take

SOME INDUSTRY CONSENSUS STANDARDS THAT PERTAIN TO EQUIPMENT USED IN FALL PROTECTION

ANSI Z359 Family of Standards

ANSI A10.32 Standard for Personal Fall Protection Equipment and Systems in Construction and Demolition

ASTM F887-05 Standard for Fall Protection in High Voltage Electrical Environments

ANSI Z133.1 Standard for Arboricultural Operations—Safety Requirements

longer to write than ANSI standards—but it can also be argued that ASTM standards may be, in the end, stronger.

FALL PROTECTION PPE VERSUS RESCUE PPE

Personal protective equipment for fall protection is generally selected as an adjunct to safety, and employers are wise to keep a sharp eye toward compliance. OSHA, as described above, drives compliance on the part of employers and on behalf of employees in the United States. When it comes to the subject of rescue, however, OSHA says little about specific equipment requirements. Nevertheless, the primary aim, as always, should continue to be for the employer to protect the worker.

At the most basic level, a rescuer should be equipped with safety gear that is at least as protective as that of the worker he or she is to rescue. This includes everything from clothing to PPE, including helmets, gloves, technical gear, and even respiratory protection. A rescuer who anticipates entering workplace hazards should take his or her initial cues from the employer's job hazard analysis, and then build on that information using whatever additional information he or she can gather.

Professional rescuers may be subject to certain standards that pertain specifically to Professional Rescue, such as NFPA 1983 Life Safety Ropes and Equipment, but these professional rescue standards are not necessarily appropriate to those performing co-worker assisted rescue. A better choice for co-worker assisted rescue personnel is for equipment to comply with ANSI Z359.4. This document is not intended for the professional rescuer, but is written specifically with the co-worker assisted rescuer in mind. Whatever equipment performance specifications the employer expects co-worker assisted rescuers to use, this information should be well documented in the rescue plan. If an employer chooses to use equipment that does not comply with the standard(s) that would most logically apply to his circumstances, he or she should be well prepared to justify that decision on the basis of safety. Additional information on selecting rescue equipment is covered in Chapter 5.

CLOTHING AND PERSONAL GEAR

Clothing worn by co-worker assisted rescuers should be appropriate to the worksite and provide at least equivalent protection to what a worker in the same environment would be expected to wear. It should provide appropriate coverage, be in good condition, and be free from rips, tears, or ragged edges to help prevent it being snagged or caught in equipment or machinery. Clothing intended to protect the worker or rescuer from specific hazards, such as heat, sparks, flame, weather, and/or wet conditions, should be selected based on meeting specific requirements for the conditions in which that employee is working on a given day.

Footwear requirements for rescuers should be considered similar to requirements for work at height. Special attention should be given to non-slip soles, ankle protection, and foot support. Stiff-soled boots help protect the foot from the effects of ladder rungs and uneven surfaces, and from "pinching" when it is squeezed between ropes or in a loop.

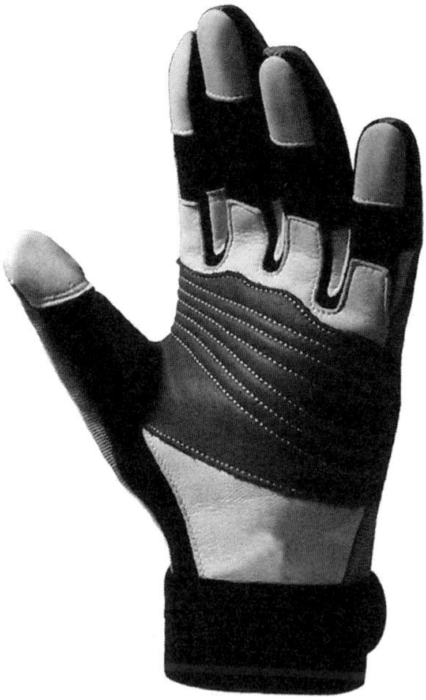

Figure 4.1 Rope handling gloves. Credit: Pigeon Mountain Industries, Inc.

Gloves for rope work are generally made of leather or synthetic leather with extra layers to protect the hand from heat and abrasion from a running rope, as shown in Figure 4.1. Finger dexterity is useful for rigging, while extra layers of protection through the palm and the groove between the thumb and fingers is desirable when handling moving ropes, such as through a brake device.

Where a variety of hazards exist, priorities must be carefully balanced. For example, a glove designed for rope work may not provide sufficient protection against other hazards, such as flame or bloodborne pathogens, while a cut-resistant glove or a glove designed to withstand flame may not provide sufficient dexterity for rope work. Where numerous hazards exist the user may need to carry more than one type of glove, or may need to prioritize and choose based on the comparative levels of potential hazard.

Harnesses

Harnesses used in fall protection and rescue are specifically designed and produced for professional use. The use of recreational harnesses, such as those used for rock climbing or caving, is generally inappropriate for professional applications, whether fall protection or rescue. Exceptions to this may include harnesses worn by professional climbers, mountain guides, and/or adventure park staff, where the hazards they face are (or mimic) the same hazards found in related recreational pursuits. Always ensure that a harness appropriately provides the type of protection needed for the employee. As a first step, simply verify that the label is marked to show it as being compliant with a standard that is applicable to the type of work in which the user is engaging.

Some examples of different types of harnesses, and the different standards that they meet, are shown in Figure 4.2, Figure 4.3, Figure 4.4, Figure 4.5, and Figure

Figure 4.2 Fall restraint/positioning harness, meets ANSI Z359.3. Credit: Heightec.

Figure 4.3 Fall arrest harness, meets ANSI Z359.1, ANSI A10.14. Credit: Heightec.

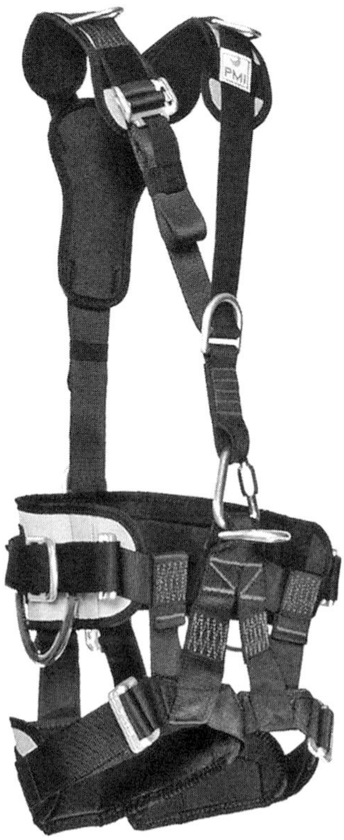

Figure 4.4 Professional rescue harness, meets NFPA 1983, ASTM. Credit: Pigeon Mountain Industries, Inc.

Figure 4.5 Arborist saddle, meets ANSI Z133.1. Credit: Heightec.

4.6. Industry standard ANSI Z359 addresses harnesses for fall protection, work positioning, and assisted rescue, while NFPA 1983 addresses harnesses for professional rescue. As this text goes to print, ANSI Z359 is in the process of moving its harness requirement criteria from Z359.1 to Z359.11. Of course, some equipment (harnesses, etc.) may be utilized across multiple disciplines, but only if the criteria for each discipline is met. Even after verifying that the harness is labeled for use in

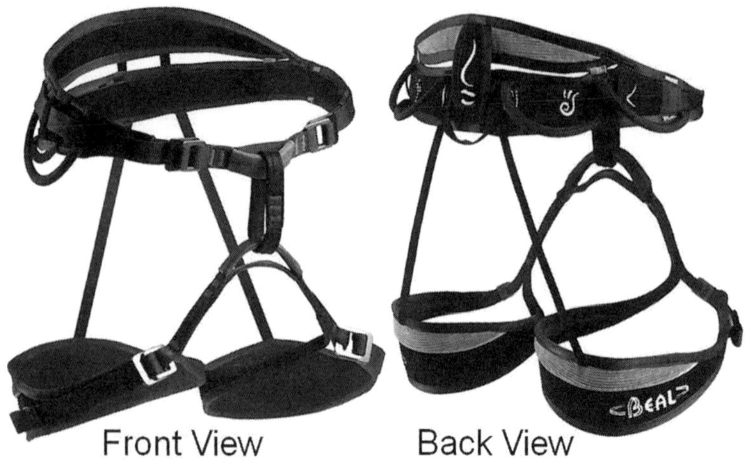

Figure 4.6 Recreational climbing harness, Meets UIAA (for recreational applications only; shown for reference). Credit: Beal.

a manner consistent with your needs, consider the design and intended use of the harness from a logical perspective, with consideration to safety.

Those engaged in co-worker assisted rescue may need to choose a harness that meets fall arrest or positioning requirements for day-to-day operations, but will probably not want to change into a different harness for rescue. In this case, an ANSI Z359 or A10 compliant harness that also meets ASTM or NFPA rescue standards is a good choice.

When selecting a harness, consider not only the appropriate standard(s) for a harness to meet but also the intended use. The intended use will also help determine what "features" you may want on the harness. Features, such as those shown in Figure 4.7, include such things as gear loops, attachment style and location, heat resistance, and material.

Harness attachment locations are a matter of safety, regulatory requirement, and personal preference. Choose a harness with a sternal attachment point to help facilitate self-rescue and co-worker assisted rescue.

Harnesses designed for fall arrest are generally of the full-body type and usually have a dorsal D-ring between the shoulder blades for connecting to the fall arrest system, as shown in Figure 4.8. When a fall occurs, the wearer is left suspended from his dorsal attachment point. It is difficult to self-rescue from this harness attachment point, and suspension for extended period of time can be quite uncomfortable. This attachment point also does not provide sufficient positioning or agility for co-worker assisted rescue, either as a subject or as a rescuer.

More recently, harnesses with a sternal attachment point (Fig. 4.9) have gained acceptance in the United States. Such harnesses have been in use throughout the world for many years, but are just now becoming common in the United States.

Although practice and some research shows that a worker can withstand a significant fall with a sternal attachment point and still be able to self-rescue, at the

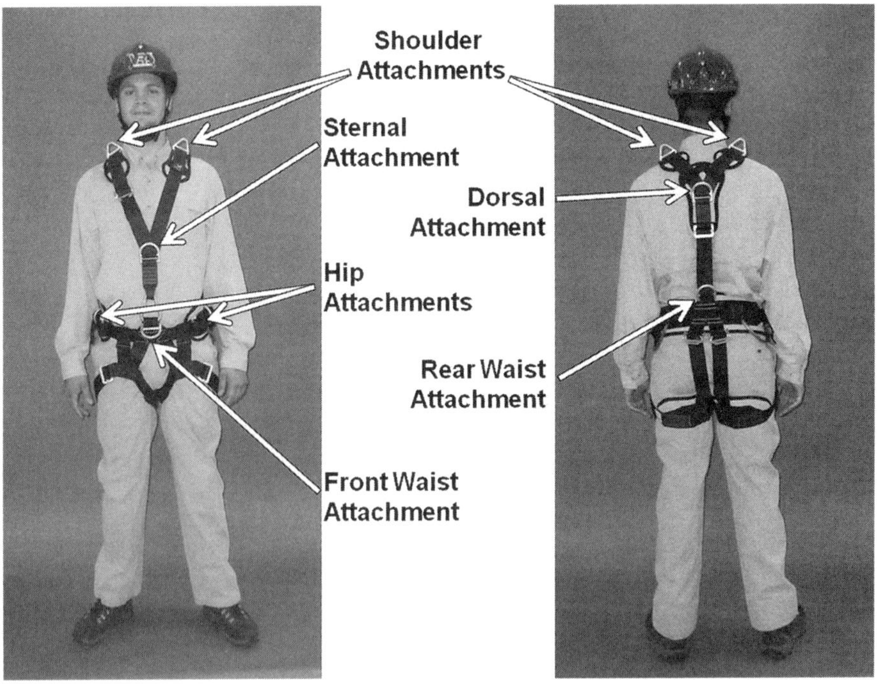

Figure 4.7 Harness features.

Figure 4.8 Suspended by dorsal D-ring.

Figure 4.9 Suspended by sternal D-ring.

moment such harnesses are restricted by ANSI standards to "limited fall arrest," meaning less than a 2 foot fall and 900 lbf impact force. Despite little evidence that limiting a fall to less than 2 feet offers any measurable advantage to the worker, this has become something of a mantra in the fall protection world.

On a practical level, the 2 foot maximum fall distance is difficult to maintain, and the gyrations of doing so may actually limit the mobility of the worker such that the fall hazard is increased. Of greater importance is the potential impact force. Over the years impact force allowances for fall protection systems and equipment have decreased. At present, the favored limit in the United States is 900 lbf. Limiting impact force and potential impact with obstructions are arguably much more important factors than limiting fall distance.

At minimum, harnesses used for fall arrest should have one designated fall arrest attachment—which may be either a dorsal or sternal attachment point, in accordance with local regulations. Fall protection harnesses meeting ANSI Z359 are *required* only to possess a dorsal D-ring, although additional attachments such as a sternal D-ring are *permitted*. Note that a frontal D-ring, at either sternal or waist level, is essential for self-rescue and/or co-worker assisted rescue. While it may be physically possible for some individuals to perform these rescue skills while wearing a harness with only a dorsal D-ring, this is neither practical nor advisable. Selecting a work harness that also has a sternal D-ring will most assuredly simplify the job. While a hard attachment point (such as a connecting D-ring) is not necessarily required from a performance perspective, some industry standards do specify this so they are quite common.

Note that with the increasing awareness of the need for the sternal attachment some older designs of harness have simply added a D-ring at this location, without re-designing the harness. While this does at least make the attachment point somewhat more accessible, it is not necessarily an ergonomic solution. In some such harnesses, being suspended from the sternal position is no more comfortable than the dorsal. Harnesses designed specifically for rescue tend to be designed so that the user sits more comfortably when suspended from waist or sternal attachments.

Harnesses designed to meet NFPA 1983 for professional rescue are required at a minimum to possess a frontal D-ring. They may or may not have additional attachment points that are designated for fall arrest, and may not be sufficient for cross-utilization in fall arrest applications. Many harnesses also have D-rings located near the left and right hip points of the waist belt as shown in Figure 4.7. These are for work positioning.

Positioning attachment points meet different criteria than fall arrest attachment points, and should not be used for fall arrest. Always read harness user instructions to determine the appropriate use and limitations of all attachment points.

Harnesses should be constructed of synthetic materials such as nylon or polyester and should be designed to support the user in a relatively upright, free-hanging position when suspended. Some harnesses, such as those designed for rope access work, are intended to be used with the worker in suspension as a matter of course, whereas a harness designed primarily for fall arrest is expected to hold the worker in suspension only in an emergency. In either case, the harness should be fitted to the worker and should be comfortable enough to be worn properly at all times. A harness should accommodate movement and be conducive to job performance.

NFPA's Standard on Fire Service Life Safety Rope and System Components Classifies Harnesses into Two Groups:

- Class II: A seat harness meant for heavy duty work by one person or in rescue situations in which another person's weight may be added in the course of the rescue.
- Class III: A full body harness meant for fall protection from falls and for use in rescues in which inversion might occur

Harnesses likely to be used for rescue generally offer a high degree of comfort to the user while suspended, whereas harnesses used exclusively for fall protection may be expected to be loaded only in an emergency and therefore may compromise suspended comfort in favor of light weight or ease of use.

In recent years a plethora of rescue equipment has been developed by companies that specialize in rescue. In the area of harnesses, particularly, great strides have been made in developing seat and full body harnesses designed specifically for rescue work. As noted previously, however, harnesses are tested to different requirements for different purposes, so identify your requirements before choosing the harness and take care to ensure that the harness you choose meets the requirements for all types of potential uses you anticipate.

Other important considerations, depending on application, may include ease of donning/doffing, compatibility with breathing apparatus, ability to switch between full body and seat-only configurations, weight, and material. A fitted, ergonomic harness that conforms to virtually any body size is the ideal solution for rescue. Thick padding is not necessarily desirable, as it can be bulky and restrictive; instead, select a harness with wide straps that distribute the weight of the wearer more comfortably across a greater area when the harness is tightened. Most importantly, pay attention to the manufacturer's specifications and instructions for the harness that you do select, and always work within those guidelines.

CHOICE OF HARNESS

Because people come in a variety of shapes and sizes, it is unusual for a single harness model to offer equal comfort to a broad range of users. An ergonomically designed harness that contours to the wearer's body and offers maximum adjustability is often a good choice. However, the best idea is usually for the employer to offer two or three different harness alternatives and allow employees to choose which fits them best. A harness that is comfortable and fits well is most likely to be used properly. Ideally, harnesses should be assigned to a single user.

Having selected the appropriate harness for your application, fitting it to the worker's weight, height, and body type is a hands-on task that can be done effectively only by placing the harness on the person who will actually wear it and physically adjusting it to fit his or her body. Because body types, and harnesses, vary significantly, employers who select two or three harnesses as "appropriate" and then allow personnel to select between them may have more success in ensuring that workers actually wear their harnesses, and that they wear them correctly.

A properly fitted harness should be snug, but not to the point of chafing or restricting blood supply. When snugly fitted, it should also be comfortable and move with the wearer. Experience indicates that a worker is disinclined to wear an uncomfortable or over-bulky harness, and may loosen or unfasten straps or not wear the harness at all. A poorly adjusted or ill-fitting harness (too loose or too tight) can cause the wearer to be released during a fall, or at the very least aggravate injury in a fall. Poor fit can also directly contribute to potentially lethal effects such as asphyxiation after a fall, and can impede circulation through either the carotid or femoral arteries.

HARNESS FIT

A well-engineered harness should support your pelvic girdle so that your weight does not create uncomfortable pressure points on the nerves and arteries in the groin and back. In a high angle rescue situation, a person may have to be hanging in the harness for a relatively long period. Look for more ergonomic padding rather than thicker padding so that the harness fits comfortably, "like a pair of jeans," and is not constricting.

Normal life expectancy for a harness is usually 3–5 years, depending on how often it is used. Some recommend that a harness be removed from service after no more than 5 years of use, regardless of condition. A worker who uses his or her harness actively may find that he or she actually retires the harness much sooner than this due to wear, but another worker who uses his or her harness less frequently may get a longer lifespan from the harness, so long as the harness passes inspection by a Competent Person. Technical reports produced by DuPont,[2] the original makers of nylon fiber, suggest that properly treated and UV-stabilized nylon will still show significant deterioration at about 10 years, even if the item is carefully stored and not used. This suggests that it may be prudent to retire any nylon product well in advance of this time frame. In the case of life safety equipment, employers would do well to take as conservative an approach as financial and training limitations will justify. If a harness manufacturer has specified a maximum lifespan in the product instructions, that lifespan should not be exceeded—although the harness may need to be retired earlier based on use or damage.

Harnesses should be marked to facilitate documentation of use and inspection. They should be inspected before each use and at periodic intervals. Marking is best achieved using a trace element control marker to write on the manufacturer's ID tags that are sewn into the harness. The ink used in pens and markers is kept in liquid form by the use of a solvent. The solvent used in some inks can damage nylon. Trace element control markers contain trace element-certified ink that meets strict requirements regarding trace element-certified content. While designed and intended to meet standards for nuclear energy, aerospace, electronic, aeronautic, and ship building industries, at this writing the content of these inks is known to be nonhazardous to nylon and polyesters commonly used in harnesses and other fall protection equipment. Take care when marking your harness to not obscure the batch number and serial number of the harness, in case the harness ever needs to be traced back to its manufacturer.

A dirty harness can be hand-washed using clear, warm (not hot) water. A mild, nondetergent cleaner may be used to remove grime, but do not use solvents or harsh cleansers. Hang dry the harness in a protected area, out of direct sunlight. Excessive heat should not be used, either in cleaning or in drying, as this can damage the fibers.

Harnesses should be stored in a cool, dry, dark location that is free from contaminants and pollutants. Avoid storing harnesses and other soft goods in the trunk of a car or a truck tool box because exposure to excessive heat and residual exhaust and chemical contaminants may cause damage or deterioration that is not readily apparent.

Seatboards may be used as an adjunct to harness comfort. The concept of a seatboard should not be confused with the old bosun's chair setup, which is a single point suspended scaffold type system originally designed to haul a sailor up the mast on a halyard to do repair work or inspect rigging. Bosun's chairs originally consisted of a wooden plank with stout lines passing underneath it and up into a bridle at chest

[2]Properties of Ropes of Dacron and DuPont Nylon, Du Pont Technical Information Bulletin X-226. Wilmington, DE, E.I. du Pont de Nemours and Co., February, 1969; Tensile Stress-Strain Properties of Fibers, DuPont Technical Information Bulletin X-82. Wilmington, DE, E.I. du Pont de Nemours and Co., May 1958; Multifiber Bulletin X-273, April 1993.

Figure 4.10 A modern bosun's chair setup. Credit: Tractel, Inc.

Figure 4.11 Seatboard, for comfort. Credit: Heightec.

level, but these days many bosun's chairs are made of nylon or other soft material. Modern bosun's chairs usually feature a body support strap or straps to help prevent the user from slipping out, as shown in Figure 4.10.

In this case it is the seat, not the straps, that provides the primary means of security to the worker, and the straps should in no way be considered the equivalent of a safety harness or fall arrest harness. A key distinction of the bosun's chair is that the seat is integral to the support of the worker; should the seat be removed, the worker's support system is totally compromised. Bosun's chairs are usually paired or integrated with a winch or block-and-tackle system to raise and lower the person in the chair.

By contrast, some rope access workers utilize a similar looking seatboard (Fig. 4.11), solely for the purpose of providing comfort while suspended. A comfort seat is not considered a "primary suspension system" but is used in conjunction with a harness-based suspension system to provide additional comfort when a worker is seated in a harness for long periods of time. In such a system, the worker's primary and secondary safety systems are attached directly to the harness. The seatboard (which may also be attached to the descender to facilitate adjustment) is only for comfort. Removal of the seatboard would not compromise the safety of the user.

Helmet

Helmets should be worn by anyone who is working at height, and by those who are working near others who are working at height. Because the consequence of even a minor bump of the head can result in more severe results when the worker must contend with additional hazards, such as working at height, wearing a helmet is an easy way to avoid problems. Helmets protect workers from bumping their own heads on obstructions, as well as from dropped objects from above.

In the United States, OSHA specifies helmets that meet the ANSI Z89.1 standard as being appropriate for industrial workers. Helmets meeting ANSI Z89.1 are further classified by *type*, which indicates the level of impact protection, and *class*, which indicates the electrical insulation level. ANSI Z89.1 Type I helmets are tested for top penetration resistance only, while Type II helmets are tested for both top penetration resistance and side impact attenuation. Electrical insulation classifications include Class E (electrical), tested to 20,000 volts; Class G (general), tested at 2,200 volts; and Class C (conductive), which are not tested to provide electrical protection. Helmet types and classifications are summarized in Table 4.1.

Unfortunately, even specifying some level of ANSI Z89.1 for a helmet to meet does not ensure that a helmet is sufficient for work at height. This is because chinstraps are inadequately addressed in that document, at least at the time of this writing. Until and unless further modifications are made to that document, there is no retention system test beyond that of a single chin strap. For maximum protection, any helmet for work at height should be fitted with a complete, three-point retention system that cradles the head, and will not permit the helmet to be dislodged from the head without release of the strap.

TABLE 4.1 Helmet Types and Classifications: ANSI Z89.1 (2009 Ed.)

	Type I	Type II
Class G (General)	Top impact protection; tested to 2200 volts	Side and lateral impact protection; tested to 2200 volts
Class E (Electrical)	Top impact protection; tested to 20,000 volts	Side and lateral impact protection; tested to 20,000 volts
Class C (Conductive)	Top impact protection; no electrical protection	Side and lateral impact protection; no electrical protection

Figure 4.12 Helmet with three-point retention system. In addition to support points on each side of the helmet, a third one is positioned at the rear. The third point helps prevent the helmet from falling forward over the eyes. The chin strap and buckle should require a significant force to cause release in order to help keep the helmet on the user's head in a tumbling fall.

To ensure adequate protection for work and rescue at height, choose a helmet that meets an appropriate industry standard, such as ANSI Z89.1, *plus* has an appropriate three-point retention system such as that shown in Figure 4.12.

Take care also to note the release force of the chin strap. Helmets meeting certain European standards have a chinstrap that releases at relatively light force. This is ostensibly to prevent a caver or confined space worker being "hung" by his helmet. However, such a light release force can also allow the helmet to be torn from the head in a fall. The safety staff responsible for your workplace should determine the most appropriate standard to cite for your application.

HELMET SELECTION

Select a helmet with a chinstrap that is mounted to at least three points. In addition to support points on each side of the helmet, a third one is positioned at the rear. The third point helps prevent the helmet from falling forward over the eyes. The chin strap and buckle should require a significant force to cause release in order to help keep the helmet on the user's head in a tumbling fall.

Ventilation holes can be a useful feature in helmets, provided the needs of the specific user do not contraindicate this. Some specifications require helmets to be flame tested and/or meet electrical conductivity tests, and for such helmets open ventilation is not appropriate. Another consideration for helmets is whether the wearer may require the use of a headlamp. Some helmets have a means for headlamp retention, while others do not. Those working in confined spaces or in dark environments will want to choose a helmet that offers this feature.

The safety professional responsible for your workplace should specify the type of helmet most appropriate to your application.

Lanyards

Most lanyards are designated for a specific purpose: fall arrest, positioning, anchorages, or the like. It's important to use a lanyard *only* for its intended purpose. Even misusing a lanyard to a lesser force than that for which it is intended, such as using a fall arrest lanyard for positioning, may compromise its subsequent performance for the intended use. Rope access, positioning, and rescue lanyards are generally intended to be used under load—and not just in an emergency.

While most lanyards are constructed of nylon and/or polyester, there is a move toward lanyards of other high-tech materials. Be familiar with advantages and disadvantages of the materials you choose, as well as with the performance characteristics of the finished product. Materials such as nylon and polyester have similar heat resistance properties, whereas ultra high modulus polyethylenes such as Spectra cannot withstand as much heat. Other fibers, such as aramids, may have higher heat resistance but are susceptible to more damage from repeated bending or other "normal" wear and tear. Table 4.2 shows some common fiber characteristics as cited by the Cordage Institute.

Lanyards, as any equipment, should be selected with the entire system in mind. How the lanyard fits together with other components to make the entire subsystem, and in turn the system, perform as a whole is of utmost importance.

TABLE 4.2 Characteristics of Some Fibers Found in Life Safety Ropes and Lanyards

	Tenacity	Melting Temp	Elongation at Break
Nylon	7.5–10.5 gpd	218–258°C	15–28%
Polyester	7–10 gpd	254–260°C	12–18%
HMPE	25–44 gpd	144–155°C	2.8–3.9%
Para-aramid	18–29 gpd	Doesn't melt; decomposes at 500°C	1.5–4.4%

Connectors

Any removable piece of hardware that connects parts of a system or subsystem together may be classified as a *connector*. For most workplace use, including onsite rescue, connectors used for attaching a user to a safety or rescue system should be of a self-closing, self-locking type.

The most common connectors found in use in workplace rescue systems are *carabiners* and *snaphooks*. Differences between these designs are shown in Figure 4.13. The most versatile type of connector for co-worker assisted rescue use is the carabiner. This is also the most commonly used type of connector in the professional

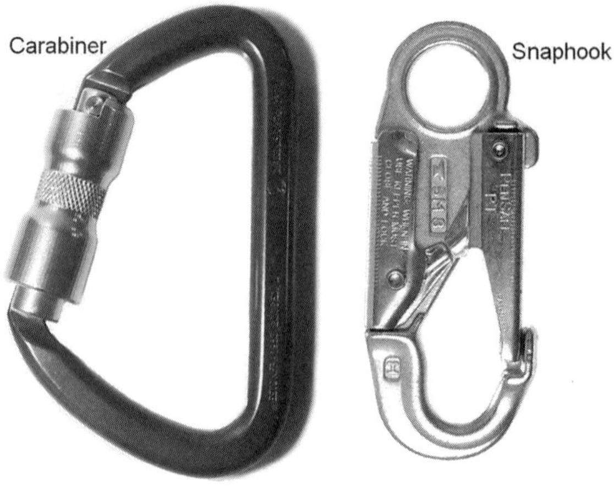

Figure 4.13 A carabiner and a snaphook.

rescue world. Carabiners selected for co-worker/industrial rescue use should meet appropriate co-worker/industrial rescue standards such as ANSI Z359.12. Carabiners meeting professional rescue standards (such as NFPA 1983) are not tested to the same requirements as those specifically designed for the workplace. These are generally appropriate only for use by professional rescuers and as a rule should not be used for co-worker assisted rescue unless they are also designated as appropriate for that purpose.

ANSI Z359 amasses all connectors used in the workplace into one single category, presumably to accommodate for their use by authorized fall protection workers who typically may have less training than rope technicians or rescuers. Connectors meeting ANSI requirements will have a minimum breaking strength of 5,000 lbs in the long axis and not less than 3,600 lbs in the short axis direction, regardless of use. These requirements apply to both snaphooks and carabiner-type connectors. Certification and performance specifications are marked on the connector itself (Fig. 4.14).

Carabiner markings should indicate the name of the manufacturer, certification organization, strength, and any standards that the carabiner meets.

Figure 4.14 Connector markings. Credit: Donald Enos/SMC.

Figure 4.15 Avoid the hazards of rollout.

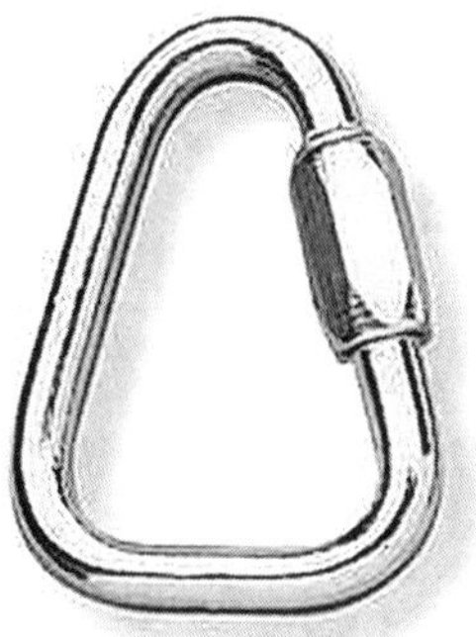

Figure 4.16 Screw link. Credit: Pigeon Mountain Industries, Inc.

For personal escape and co-worker assisted rescue, select carabiners that are rated to at least 22 kN (5,000 lbs) in the long axis and to 16 kN (3,600 lbs) in the short axis, and that feature a self-closing, self-locking method of closure. Connectors must also be compatible with the connection to which they will be attached.

Compatibility is a concept that is based on relationships between multiple components. It is not possible to select one carabiner that will always be compatible with every possible connection, so the competent user must exercise judgment about this in the field. Shape, size, and gate design should allow the carabiner to maintain position in its connection without jamming, loosening the connection, or being susceptible to *rollout*. Rollout, demonstrated in Figure 4.15, is the phenomenon that occurs when two pieces of hardware twist against one another in such a way so as to cause inadvertent release of one of the connectors, often by breakage.

Another type of connector, known as a "screw link," shown in Figure 4.16, was once popular in safety rigging and rescue. Neither OSHA nor ANSI Z359 make accommodation for these in their standards, at least at the time of this writing. For this reason, the appropriateness of screw links for use in general industry is questionable. While experienced, competent users may be quite capable of using screw links safely, and will recognize certain benefits in using them, from a regulatory perspective there is presently little justification for their use in industry.

> Screw links are sometimes used as a harness buckle. This is an acceptable use of a screw link, as long as the manufacturer has designed the harness with this in mind.

The integration of connectors into system rigging is discussed in further detail in Chapter 5.

Braking Devices

A *braking device* is a piece of equipment that is used to impart friction on a rope for the purpose of slowing descent. Friction is the primary means of limiting or preventing a rope's ability to travel. In both high angle and low angle rescue systems, friction is used to control how quickly the load descends. For rescue purposes, the load may consist of a patient, rescuer(s), litter, or some combination thereof.

Braking devices may be used for descending, that is, traveling down a rope, or for lowering a load from above. To avoid confusion, for discussion purposes it is helpful to delineate the use of the term *descender* as being in reference to a device used by an individual to control his or her own descent as he or she travels down the rope, and to use the term *lowering device* to refer to a device that is used in a stationary configuration to lower a load from above. Some braking devices may be intended for one or the other of these types of uses, but most braking devices may be used for both purposes.

Braking devices are used in rope access work and in various types of rescue, including co-worker assisted rescue and professional rescue techniques. Braking devices must be of a design that is compatible with other components of the system in which they are used—especially the host rope. Some braking devices are designed for use with one specific type or design of rope. In such cases the information will be called out in the user instructions, and possibly even on the device itself. Most braking devices are intended for use with a variety or range of ropes within specified diameters. Even braking devices that are not designated by the manufacturer as being rope-specific will vary in performance from rope to rope depending on everything from diameter to construction to material. It is left to the Competent Person to make an educated decision regarding the compatibility and use of braking devices based on a combination of product instructions, education, and experience.

According to ANSI Z359 requirements for assisted rescue, any descender used for assisted rescue must lock off if the control mechanism is released or squeezed too hard, must be easily threaded, and must have a controlled rate of descent. Professional rescue (NFPA) standards do not have this limitation, so as to allow greater versatility and breadth to the user. In most cases, braking devices that do not feature the automatic locking mechanisms specified by ANSI Z359.4 should not be used for co-worker assisted rescue, even though they may be quite appropriate for professional rescue. For co-worker assisted rescue, always select a descender that meets ANSI Z359.4 requirements, or an equivalent.

Before using a descender at height, always practice using it on a level or gently sloping surface in a controlled environment. Be aware that descenders used for rope access in everyday work may not meet the standards required by ANSI Z359 for co-worker assisted rescue.

Rope Grabs

A *rope grab* is a device that, by means of squeezing or compressing a cam against a rope, moves easily along the rope in one direction but grips the rope tightly when pulled in the opposite direction. Rope grabs are used for a variety of purposes in both work and rescue applications. Different types of rope grabs are appropriate for different uses.

Rope grabs that are appropriate for fall arrest (*fall arresters*), such as that shown in Figure 4.17, will meet an appropriate standard for fall arrest, and should be labeled as such. Always be sure that the fall arrester you are using meets the requirements of the country/jurisdiction where work is taking place. The idea behind a fall arrester is that it will travel up the rope easily, perhaps even trailing, as the user moves upward, but if the person falls the fall arrester will engage, gripping the rope with enough force to stop the fall without damaging the rope. A certain amount of "slip" may be designed into the fall arrester for this purpose.

It is not unusual for a fall arrester to be used as a primary safety connection for a user or users, so in most cases a rope grab should be of a design that cannot easily be inadvertently removed from the rope. To every rule there is an exception, and in this case the exception is the backup devices used for rope access. Due to the unique nature of rope access work, backup devices used in this field must be relatively easy to put on and take off of a rope and of a design that can be installed onto a horizontal or oddly positioned rope so as to facilitate transitions and rope-to-rope

Figure 4.17 A fall arrest rope grab.
Credit: Pigeon Mountain Industries, Inc.

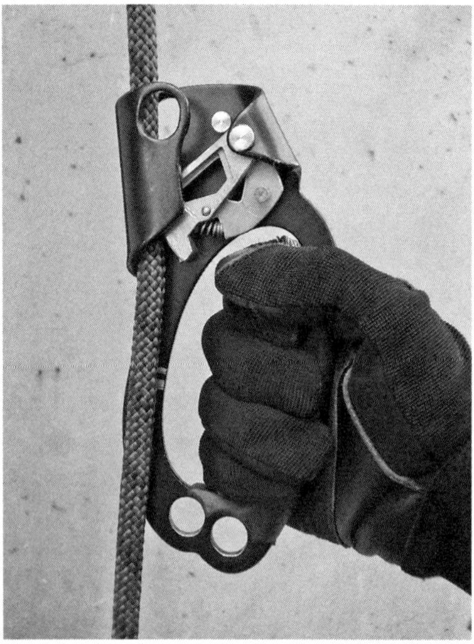

Figure 4.18 Personal ascender.

transfers, yet still not be susceptible to inadvertent removal. Further, the user must consider potential for a two-person load, such as during rescue. For these reasons, the user may elect to use for this application devices that are not necessarily certified to typical "fall arrest" standards.

Rope grabs specifically designed for gripping a rope for the purpose of ascending it are generally known as *personal ascenders*. These are usually designed with a handle, for gripping, or with connection points for rigging into a harness or ascending system, as demonstrated in Figure 4.18.

Personal ascenders are intended for ascending with only one person's body weight, not for hauling systems in which forces are multiplied. Some of the unique characteristics of ascenders include a handle for gripping, a more aggressive camming device, and a design that makes the unit easy to place on and take off of the rope. Ascenders are usually not designed to withstand a fall, and are not recommended as a primary safety attachment. They tend to be very unforgiving in the event of misuse, so personal ascenders should be used with great care and excessive loading must be avoided.

The primary standard governing personal ascenders is UIAA, but this is a recreational standard and may not be appropriate for industrial/rescue applications. ANSI Z359 does not address personal ascenders, but NFPA 1983 does specify some test methods for personal ascenders that are more substantial than the UIAA recreational requirements. These may be a good alternative for the industrial user to consider.

Rope grabs designed for use in rigging and hauling (Fig. 4.19), such as in rescue systems, perform differently from fall arresters, rope access backup devices,

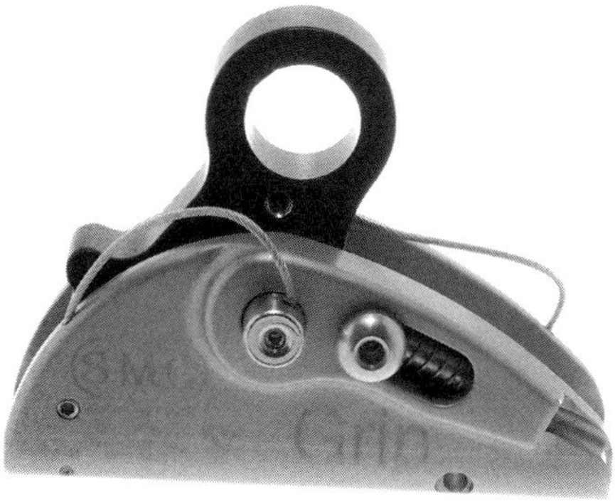

Figure 4.19 Rope grab designed for rigging and hauling. Credit: Pigeon Mountain Industries, Inc.

or personal ascenders. Neither as forgiving as a fall arrester nor as aggressive as a personal ascender, rope grabs for rigging are intended to be used as a part of a hauling system or other rigging subsystem specifically for the purpose of gripping the rope.

This type of rope grab is quite useful in co-worker assisted rescue, so understanding the capabilities and limitations of this equipment is essential for potential responders.

Personal Escape Systems

Any person working at height, whether engaged in fall protection techniques, rescue, or some other system, should be equipped with emergency egress capability. Most common are compact, prepackaged personal rope descent systems that can be carried in a pocket or small bag (Fig. 4.20). Normally a personal escape kit will contain some combination of rope, descender, and connector(s).

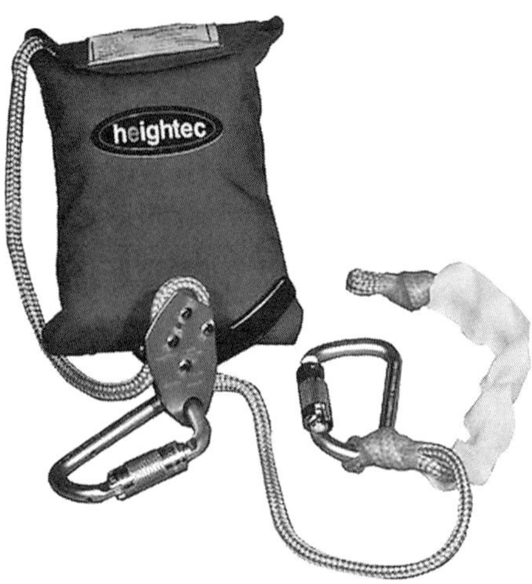

Figure 4.20 Pocket-sized escape system. Credit: Heightec.

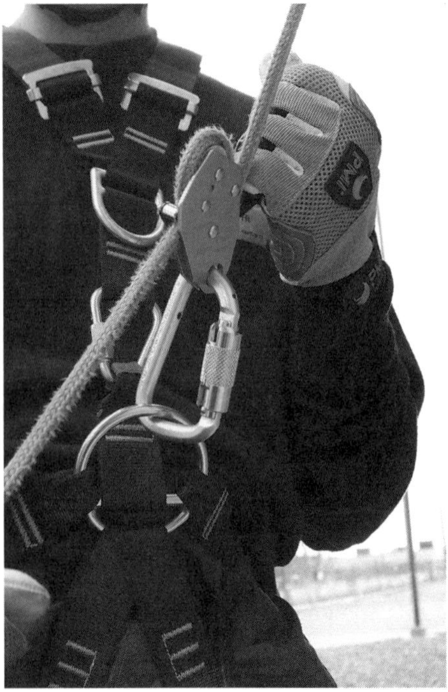

Figure 4.21 Autolocking escape descender. Credit: Pigeon Mountain Industries, Inc.

Small diameter ropes are often used for personal escape, either of nylon or aramid fibers. Nylon ropes are desirable for their force absorption characteristics and reliable performance. While aramid's resistance to high temperature may sound attractive for those working in high heat environments, great care should be taken with aramid ropes as this fiber has virtually no shock absorption capability. Aramid is also a relatively brittle fiber that is especially prone to self-destructive internal damage under repeated bending abuse such as might occur over time when stored in a pocket or bag. The amount of benefit offered by such fibers should be quantified in terms of seconds gained at likely/survivable temperatures, and then weighed against the cost and potential hazards.

Small diameter ropes used for egress require a special descender with sufficient friction to perform over the smaller surface area of ropes. Philosophies of escape descenders tend to fall into one of two camps: *autolocking* (Fig. 4.21) or *free running* (Fig. 4.22).

The autolocking type of Personal Escape descender offers greater security against an uncontrolled descent, but could be prone to being locked off unintentionally during escape. Autolocking types are also usually more difficult to put on and take off of the rope—which may be considered as an advantage or a disadvantage, depending on the situation. Free-running escape devices, such as the Escape 8 or the Personal Escape Device (PED), may be more desirable in particularly hazardous atmospheres where a timely escape is of the essence and/or where becoming stuck on rope could be catastrophic, such as in running water, or fire.

Figure 4.22 Free running escape descender. Credit: Pigeon Mountain Industries, Inc.

SUMMARY

Personal protective equipment should be selected based on appropriateness to the task to be performed, as well as meeting regulatory requirements in the jurisdiction where work will take place. Attention to safety is always paramount. PPE does not stop with harnesses and helmets, but includes everything from the clothing one wears to connecting elements.

Not all PPE is appropriate to every use. The user should consider fit, desired performance, and other aspects of safety. Special consideration should be given to the appropriateness of equipment for the specific environment in which it is being used.

POST-CHAPTER QUIZ

Chapter 4

1. PPE generally refers to items
   - a) Used to perform a task while working, such as tools or specialized equipment
   - b) Typically worn by a worker to provide protection against recognized hazards
2. Rescue equipment refers to items
   - a) Used to perform a task while working, such as tools or specialized equipment
   - b) Typically worn by a worker to provide protection against recognized hazards
3. True or False: If a harness is appropriate for fall protection, you can be sure it will be appropriate for rescue.
4. True or False: OSHA sets forth certification requirements for harnesses.
5. The purpose of ANSI in standards development is to
   - a) Write documents for all fall protection equipment in the United States
   - b) Verify procedural conformity for industry consensus standards
   - c) Provide a bridge between European standards and standards in the United States
   - d) Become the International Standards Organization

6. Choosing a harness with a ______________ attachment point will help facilitate self-rescue and co-worker assisted rescue techniques.

    a) Dorsal
    b) Waist
    c) Sternal
    d) Shoulder

7. True or False: Handled ascenders may not be used for fall arrest.
8. The phenomenon that occurs when two pieces of hardware twist against one another in such a way so as to cause inadvertent release of one of them is called

    a) Breakout
    b) Rollover
    c) Rollout
    d) Breakover

9. Any piece of hardware that connects parts of a system or subsystem together may be classified as a

    a) Carabiner
    b) Twistlock
    c) Connector
    d) Descender

10. The ____________________ type of personal escape descender offers security against an uncontrolled rappel, but could be prone to being locked off unintentionally during escape.

    a) Autolocking
    b) Free-running

ANSWER KEY

1. b
2. a
3. False
4. False
5. b
6. c
7. True
8. c
9. c
10. a

CHAPTER 5

RESCUE EQUIPMENT AND SYSTEMS

By the end of this chapter you should understand:

- The distinction between "life safety equipment" and "commodity equipment."
- Component considerations for technical rescue systems.
- How to select key items of hardware (connectors, rope adjusters, etc.).
- How to select rescue rope, slings, and webbing.
- Different considerations for patient transport equipment.
- The purpose of a pre-engineered rescue/retrieval system.
- How to select a pre-engineered rescue/retrieval system.

In Chapter 4 we explored personal equipment for co-worker assisted rescue responders. That chapter focused primarily on equipment that is worn or used directly by the person. This chapter will expand on that concept, adding in equipment that is intended for use in the actual rescue system.

Every employer is faced with the challenge of how to maintain a co-worker assisted rescue capability that will ensure readiness for the types of responses that may be encountered, without sinking excessive amounts of time and money into co-worker assisted rescue equipment and training. Providing too little information/equipment will leave responders ill-equipped, while providing too much information/equipment is likely to result in responders not having the specific equipment they need available when they need it, or of being so overwhelmed with information that they forget how to use it in an emergency.

The solution rests in understanding what potential incidents might occur, and in judiciously equipping responders with the correct resources and training—neither too little nor too much.

Rescue equipment requirements may vary based on the worksite, potential incident type, and environment. Either personal escape or co-worker assisted rescue from height might be necessary on a tower, from a catwalk, beneath a roadway, off a bridge, on the side of a building, in the bowels of a ship, over a cliff, in a cave, in the rubble of a natural disaster, or in any number of situations. While the basic

Falls from Height: A Guide to Rescue Planning, First Edition. Loui McCurley.

principles of gravity and rigging ropes for rescue purposes remain the same in any of these environments, procedures for applying those principles certainly must change to accommodate circumstances. In addition, specific types of equipment may be advantageous—or even critical—to consider in one environment as compared with another.

CONSIDERATIONS IN SELECTING EQUIPMENT

Selection of appropriate rescue equipment requires technical understanding as well as a working knowledge of regulatory requirements. Co-worker assisted rescue responders are not excused from fall protection requirements, and if they are exposed to a potential fall during rescue they should be protected accordingly. That said, when it comes to the rescue system itself, regulatory requirements for rescue equipment differ from requirements for fall protection gear. This is largely because rescue invariably occurs from locations where a person was not actually intended to be. Because it can be difficult to standardize expectations of what kinds of rescues might be required, requirements must remain flexible. For this reason, technical knowledge and experience are critical factors in planning successfully for rescue needs.

To state what should be glaringly obvious: only equipment expressly designed for life safety use should be used for work, personal escape, or rescue at height. While this should seem rather obvious to most safety professionals, it bears specifying especially to those who are familiar with the lifting industry. Some may be tempted, perhaps for reasons of convenience, to employ certain "utility" lifting products in the safety system. Ironically, this is most likely at the most basic elements of the chain—in components such as anchorages, rope, and connectors.

The differences between "life-safety equipment" and equipment designed for "commodity use" can be easy for the uninitiated to overlook because (aside from certification labels) these differences are usually invisible to the naked eye. Characteristics such as strength, deformation, efficiency, and compatibility with other equipment are specified more closely and measured with greater accuracy and confidence for life safety equipment than for equipment designed for the commodity market.

Rescue equipment must be selected and utilized in a manner that is specific to its intended use. Safety planners, as well as at-height workers, should take it upon themselves to confirm that equipment selected does, indeed, meet the requirements necessary (both performance and regulatory) for the intended application. In general, equipment designed for rescue is stronger and more conducive to extended use under load than equipment designed for personal use.

There is more than one industry standard that offers specifications for rescue equipment. While the ANSI Z359 family of fall protection standards, published by the American Society of Safety Engineers,[1] pertain primarily to personal protective equipment, such as that described in Chapter 4 of this book, one document within this family (ANSI Z359.4 Safety Requirements for Assisted Rescue and Self-rescue

[1]American Society of Safety Engineers, 1800 E. Oakton Street, Des Plaines, IL, 60018.

Systems, Subsystems, and Components) does offer guidance for equipment that is intended for co-worker assisted rescue.

ANSI Z359.4 was first published in 2007, and a 2013 revision has just been published as this book goes to print. This document specifies minimum design and test requirements for equipment used in rescue and retrieval of workers after a fall. This document is not intended to specify equipment for professional rescuers, but for pre-planned, co-worker assisted rescue operations. Equipment for one-person rescue systems (130–310 lb) and two-person rescue systems (160–600 lb) is specified. In addition, connectors must meet the requirements of ANSI Z359.1.

At the other end of the spectrum is NFPA 1983,[2] developed by the National Fire Protection Association expressly for fire service personnel and operations. This standard is not directed at co-worker assisted rescue applications, but employers and safety professionals may wish to consider equipment that meets this standard where equivalent alternatives designed specifically for the workplace are not available.

Finally, there are several ASTM[3] documents available through the ASTM F32 committee on search and rescue that may be of use to the industrial safety/rescue professional in certain applications. The ASTM rescue standards are not specific to workplace rescue nor to any particular branch of emergency services.

SOME INDUSTRY CONSENSUS STANDARDS THAT PERTAIN TO EQUIPMENT USED IN RESCUE INCLUDE:

ANSI Z359.4: Safety Requirements for Assisted-Rescue and Self-Rescue Systems, Subsystems and Components

ASTM F1764: Standard Guide for Selection of Hardline Communication Systems for Confined-Space Rescue

ASTM F1956: Standard Specification for Rescue Carabiners

ASTM F2116: Standard Specification for Low Stretch and Static Kernmantle Life Safety Rope

ASTM F2436: Standard Test Method for Measuring the Performance of Synthetic Rope Rescue Belay Systems Using a Drop Test

NFPA 1983: Standard on Life Safety Rope and Equipment for Emergency Services

Because it is reasonably foreseeable that rescue loads may comprise more than one body weight, manufacturers may use a higher *design load* for rescue gear than for fall protection or personal escape equipment. Conversely, because rescue techniques should eliminate potential for long falls, rescue equipment generally need not be rated for high *impact loads* such as might occur in a fall.

Connectors, descenders, and other hardware used by professional rescuers may have fewer fail safe mechanisms built in than similar equipment used for personal escape or co-worker assisted rescue by at-height workers. This is to facilitate versatility and ease of use, and is justifiable based on the fact that professional rescuers

[2]National Fire Protection Association, NFPA, 1 Batterymarch Park, Quincy, MA, 02169-7471.

[3]ASTM International, 100 Barr Harbor Drive, P.O. Box C700, West Conshohocken, PA, 19428-2959.

are more highly trained in the use of this equipment, their experience in rescue methods and systems is greater, and the purpose of their using the equipment is specifically for rescue. In the case of at-height workers, their safety system is secondary to their primary purpose (which may be maintenance, construction, services, etc.), versatility is less of a concern than consistency, and their levels of experience with the rescue systems may be limited.

Rescue equipment designed and intended for use by professional rescuers is generally not recommended for use in co-worker/industrial assisted rescue unless the workers being equipped are thoroughly and appropriately trained and experienced, such as might be the case in a situation involving a standby team of professional rescuers on an industrial site.

There are certain notable differences between equipment used for assisted rescue in the workplace as compared with equipment used by professional rescuers to employ advanced rescue techniques. The primary focus of this text is the equipment and techniques used for co-worker assisted rescue; however, this book is not a substitute for practical training and experience. The reader is cautioned to secure hands-on training specific to the equipment and methods he or she will use, and to use only that equipment for which he or she is trained.

Equipment Strength

All equipment used for rescue should be able to withstand and perform effectively with whatever loads might be applied. This may include more than one body weight at a time, and may be intensified by friction and rigging angles. Rescuers must be able to trust their equipment at all times, without question. To this end, rescue gear should be kept separate from gear used for everyday work, and should be maintained in a ready state. This helps to ensure that when a rescue happens, all components are available, fit for service, and not worn out.

Personal escape and co-worker assisted rescue equipment and systems should excel in simplicity and ease of use. Equipment used for advanced rescue techniques employed by professional rescuers tends to be quite versatile and multifunctional in nature than that used for co-worker assisted rescue, since workers may not have opportunity to practice personal escape and “buddy rescue” nearly as often as professional rescuers might train for their specialty.

Safety Margins

There is an important distinction in the way that breaking strengths for life safety ropes and equipment are reported, and this distinction often carries over to other equipment as well. Those who are accustomed to working with cranes and lifting equipment may be surprised by an apparent lack of *safe working load (SWL)* information provided by life safety rope and equipment manufacturers. This is not an oversight, but is intentional.

Safe working load is a figure that is commonly derived as a percentage of breaking strength and applied to lifting equipment in non-life safety applications. There is no standard for specification of safe working loads in life safety equipment, nor the relationship between SWL and *minimum breaking strength (MBS)*. The problem with SWL is twofold. To begin with, the SWL does not give the user sufficient information to take into account the relationship between components (for example, rope and other equipment) as they are used together. Second, knowing only the SWL of several pieces of equipment does not give the user the information he or she needs to determine the overall safety factor of the equipment that is being used when it is rigged as a system—which is really the most important thing for him or her to know. For this reason, life safety equipment standards usually specify that MBS be provided for equipment intended for life safety use.

The minimum breaking strength of a component that is intended for life safety use is generally specified as three standard deviations below the mean breaking strength of a minimum number of samples of that component when tested in manner of use. The testing and reporting methods specified help to ensure a high confidence in the cited MBS, which gives the rescuer a piece of information he or she can really use. With this information, the rescuer can apply other known information to determine a *system safety factor*.

A system safety factor is calculated as the ratio between the weakest link in a system and the load that is expected to be applied. As a rule, the higher the probability and the consequence of failure, the higher the system safety factor should be. A user has to know what the actual expected breaking strength of a component is in order to effectively calculate system safety factors.

ROPES FOR RESCUE

The foundation of nearly every high angle personal-escape or co-worker assisted rescue system is rope. Only ropes designated as being intended for life safety use should actually be used to support human life. According to the Cordage Institute,[4] an organization of rope manufacturers in the United States, life safety ropes have unique specifications and characteristics, as compared with commodity ropes. For example, commodity rope diameters are often measured based on *mass per unit length* with results cited as *nominal diameter*, whereas life safety rope diameter is measured more precisely and cited as *actual diameter.* Using actual diameter measurements helps ensure compatibility between ropes and auxiliary equipment. Measurements for breaking strength, elongation, and other characteristics also vary for life safety rope as compared with commodity rope.

> Life safety ropes in the United States should meet life safety-specific standards such as CI 1801 or CI 1805.

For these reasons, the Cordage Institute offers separate, specific standards and test methods for different types of life safety ropes. These test methods and reporting criteria are often cited by other, industry-specific organizations such as NFPA, ANSI, ASTM, and others, as they develop user specifications.

[4]Cordage Institute, 994 Old Eagle School Road, Suite 1019, Wayne, PA, 19087.

Dynamic rope, such as that which meets the criteria of appropriate UIAA or similar standards, is generally used where an individual must climb above an anchor point—known in the vernacular as *lead climbing*. Lead climbing is most common among recreational rope users, but is a valuable technique for rescue technicians who must access a victim from below. Lead climbing techniques are touched upon in Chapter 6. This is a highly specialized skill and should only be used by specially trained and experienced rescue technicians. The primary limitation of dynamic rope is that it elongates significantly under load, making it very difficult to control effectively in lowering and raising operations.

Low stretch ropes and *static ropes*, as defined by Cordage Institute standards, provide greater stability for lowering and raising. Low stretch ropes do offer a limited amount of force absorption, but are not appropriate for taking significant falls. This elongation also allows a rope to stretch somewhat under load, but offers a compromise between the dynamic Ropes described above and their more structurally unyielding counterpart, static ropes. Static ropes provide the maximum amount of precision and stability for lowering and raising loads. Static ropes are defined by the Cordage Institute as having less than 6% elongation at 10% of minimum breaking strength. Static ropes all but eliminate the inherent *creep* in rescue systems, making patient loading and transport a more predictable and controlled operation.

There are several important factors in selecting life safety rope that are largely determined by technique and site considerations. What is the range of loads that will be put on the rope? What is the potential fall distance? How critical is precise control of the load? Weigh these factors with your own specific criteria to select the right rope for the job.

Life safety ropes are usually of a kernmantle construction. Rope construction helps determine rope performance characteristics. In the past, a type of rope known as "laid rope" was very common. Laid rope consists of three or more bundles of fiber which are simply twisted together in barber-pole fashion. These ropes are generally of the cheaper varieties, and have historically been common in fall arrest systems. While some such ropes are still in limited use for life safety today, laid rope is more vulnerable to abrasion, which decreases total rope strength. Laid rope can also be more likely to get caught on things and rip or tear, and offers a great deal of mechanical deformation and elongation in use, making precise movement and stabilization very difficult and causing "spin" when loaded.

Solid-braid ropes are another type of rope that have historically been commonly used for fall arrest; solid-braids are very flexible, offering a nice "feel." This flexibility can make knots very difficult to untie, however, and can increase the ropes' susceptibility to wear. The strength of a solid-braid rope is dependent on all of the strands; as with a laid rope, this means that the overall strength is immediately affected by any wear, cuts, or damage.

The best rope by far for life safety use is the type with a kernmantle construction. The term "kernmantle" refers to the concept of a "core and sheath" design wherein the primary load-bearing fibers form an interior "core" and a protected braided sheath covers the core. Although kernmantle rope is generally less flexible and more expensive than laid or braided rope, it is more resilient and usually offers a longer service life.

TABLE 5.1 Fiber Performance Characteristics

	Breaking Tenacity	Chemical Susceptibility	Melting Temp	Elongation at Break
Manila	5–6 gpd	Acids, alkalis	Chars at 148°C	10–12%
Sisal	4–5 gpd	Acids, seawater	Chars at 148°C	10–12%
Nylon	7.5–10.5 gpd	Mineral acids	218–258°C	15–28%
Polyester	7–10 gpd	Sulfuric acids, alkalis	254–260°C	12–18%
HMPE	25–44 gpd	Minimal	144–155°C	2.8–3.9%
Para-aramid	18–29 gpd	Strong acids and bases	Decomposes at 500°C	1.5–4.4%

The fiber used in the construction of a rope also helps to determine performance characteristics of the end product. A comparative reference of fiber characteristics, such as those outlined in Table 5.1, can be found in Cordage Institute's CI 2003 document. Manila and sisal used to be common rope fibers, but this is no longer the case due to strength and longevity issues. In fact, OSHA now specifically states in some rules that ropes and straps used in lanyards and lifelines are to be made from synthetic fibers.[5]

Although polypropylene can still be found in some vertical lifeline fall protection systems, this fiber is also a poor choice for life safety. Not only does polypropylene degrade relatively quickly, especially when exposed to ultraviolet light, it is also not a very strong fiber. All other things being equal, a polypropylene rope must be thicker than a polyester or nylon rope to achieve equivalent strength—a fact that also has implications for compatibility of equipment.

Today, most life safety ropes are of either a nylon or a polyester construction, or some combination of the two. These two fibers have similar strength, elongation, heat, and durability characteristics, and the end results of each usually perform quite similarly. At a strength of approximately 7–8 grams per denier, it takes only an 11 mm (7/16 in) nylon or polyester rope to achieve in excess of 6,000 lbf strength.

In some markets there is a move toward what are considered "high tenacity fibers," such as high modulus polyethylenes (HMPE) or para-aramids. More expensive does not necessarily mean "better" however, so caution should be used when selecting products of these fibers. Although initial strength may be higher, many of these fibers do not perform well in the areas of force absorption, heat resistance, or knotability.

Rope used for personal escape, co-worker assisted rescue, and/or professional rescue should be specifically designed and intended for life safety use. Commodity and utility ropes purchased at hardware and general purpose stores are not tested and certified to as exacting criteria and safety margins as are ropes designed for life safety.

Life safety rope is made for specific purposes and applications. Rock climbers have different priorities that mountaineers, who in turn have different priorities from

[5]29 CFR 1926.502(d)(14); 1910.66 App C.

cavers, who prioritize differently yet from rescue personnel. Interestingly, the requirements of each of these categories of life safety rope users tend to revolve around similar performance considerations—it's just that they all want a different combination of those characteristics.

Life safety rope is generally classified into three types: dynamic, static, and low stretch. Each of these three types of rope are tested to different standards and criteria. From a use perspective, one criterion life safety rope users use to select rope is based on how far they anticipate potentially falling in a worst case scenario. The fall distance is most important as it relates to the amount of rope available to catch the fall. Together, these measurements are used to calculate a fall factor (Fig. 5.1).

TYPES OF ROPE

- Dynamic: Rope that is designed to absorb force during a fall. Used primarily by lead climbers, and those who anticipate significant fall potential.
- Static: Rope that has less than 6% elongation at 10% MBS. Used by those who want minimal stretch and/or "bounce" under load.
- Low stretch: Rope that has 6–10% elongation at 10% MBS. Used by static rope users who want a little more force absorption in their systems.

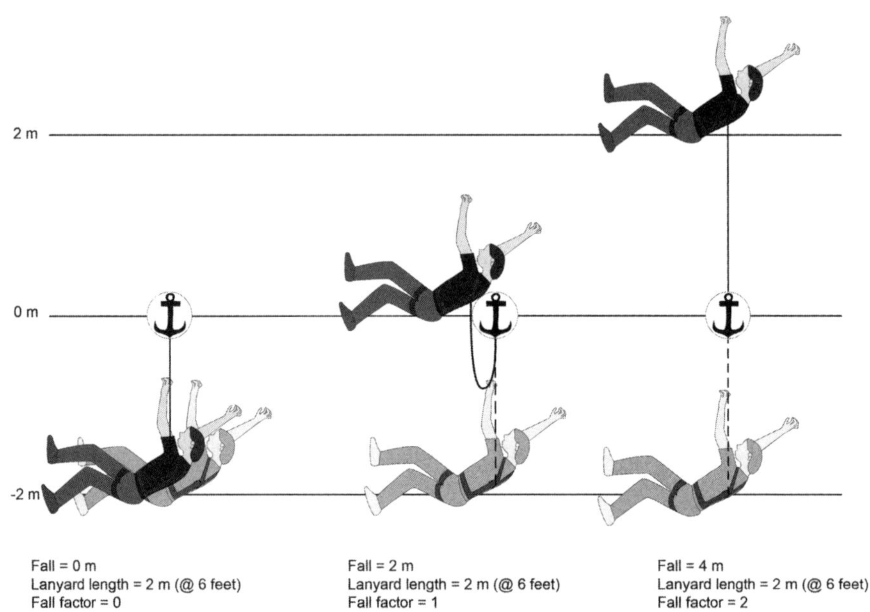

Figure 5.1 Fall factor. Credit: Pigeon Mountain Industries, Inc.

Measuring Rope Performance

Most people would prioritize characteristics of life safety rope, for whatever purpose, to include:

- Strength
- Diameter
- Force absorption
- Elongation
- Abrasion resistance
- Compatibility with other equipment
- Hand

We will explore each of these characteristics in the text to follow, but for additional detail regarding how these are measured consult an appropriate standard/test method such as CI 1801.

Strength Strength requirements for life safety rope are most important to those who are using the rope in raising, lowering, ascending, or rappelling. While most rope consumers never come anywhere close to pushing the strength limits of the equipment they use, a necessary margin of safety is factored into any system. In some extreme environments and uses, such as rescue and industrial, heavy loads and complex systems combined with safety margin requirements can create a challenge. Acceptable safety factors range from 4:1 to 10:1, and these are most appropriately applied to the completed system, not just the individual components such as rope. This not only requires a fairly sophisticated end user to calculate, but also requires quite a high starting strength in order to compensate for strength reductions as the equipment is integrated into a system. According to Cordage Institute specifications, static and low stretch rope must meet minimum strength requirements as shown in Table 5.2.

Diameter Most life safety ropes range in diameter from 7.5 to 13 mm. Accurate assessment and reporting of diameter is critical for life safety ropes. Most of the auxiliary equipment designed for use with life safety ropes works with very specific

TABLE 5.2 Minimum Breaking Strength (MBS) by Size (diameter), as Specified by Cordage Institute CI1801

9/32″	(7 mm)	2,200 lbf	(9.8 kN)
5/16″	(8 mm)	2,875 lbf	(12.8 kN)
3/8″	(10 mm)	4,500 lbf	(20.0 kN)
7/16″	(11 mm)	6,000 lbf	(26.7 kN)
1/2″	(12.5 mm)	9,000 lbf	(40.0 kN)
5/8″	(16 mm)	12,500 lbf	(55.6 kN)

rope sizes. Friction and gripability are important considerations, but so is weight. Balance of these factors is largely a matter of personal preference and compatibility with other equipment.

Force Absorption Impact force is an especially important consideration for those who are lead climbing above their protection, thereby opening themselves to a high fall factor with significant impact potential.

Ropes used for this purpose are required to absorb energy during a fall so as to reduce the force transmitted to the climber and anchorage(s). In the case of static and low stretch ropes, impact force is not generally considered an important factor. These types of ropes are common in personal escape and rescue systems, where significant falls are mitigated, and in fall arrest systems, where force is absorbed by other components such as the force-absorbing lanyard and/or rope grab.

Elongation Ropes that offer high impact force absorption generally also have a great deal of elongation. Elongation poses a practical concern where loads are being raised, lowered, and positioned on a vertical plane, such as commonly occurs in rescue. Ropes with too much elongation can require more effort to raise, can "bounce" the load, and can cause a stopped load to creep dangerously. For this reason, ropes with lower elongation are preferred for raising, lowering, and positioning heavy loads.

Abrasion Resistance Abrasion resistant ropes offer greater security on rough surfaces such as steel, concrete, and other industrial settings. Because ropes with the best abrasion resistance generally do not also boast the softest hand (see below), experienced rope users are often evidenced by the fact that they select ropes with a tighter sheath weave and stiffer characteristics.

Compatibility Auxiliary equipment selected for use with life safety rope should be selected specifically according to purpose and with consideration to the specific rope to be used. Rope construction is important with some devices, as is sheath material, flexibility (too much or too little), and even sheath slippage.

Hand The term "hand" is used to describe, literally, the feel of a rope in one's hand. A rope must be manageable and easy to work with, but these terms are really very subjective. An experienced user will have different priorities than an inexperienced user. The soft hand and flexibility that most inexperienced rope users prefer are easily recognized by seasoned veterans as indicative of poor abrasion resistance, sheath slippage, and inefficiency in raising and lowering systems.

Quality Assurance In addition to the technical considerations involved with the manufacture of life safety rope, quality must be considered a key factor. Some user groups of life safety rope now mandate that qualifying manufacturers meet specific quality assurance criteria, such as certification to an ISO quality standard.

PMI Rope Inspection Log

Purchased From | Purchase Date / / | PO Number | User's Serial or ID Number | I.D. Markings

Manufacturer's Lot Number | Date In Service / / | Assigned Use | Diameter/Length | Color

INSPECT ROPE, CONNECTORS, AND ENDS FOR DAMAGE OR EXCESSIVE WEAR BEFORE AND AFTER EACH USE. IMMEDIATELY RETIRE ALL SUSPECT ROPES.

Date Used	Location	Type of Use	Exposure	Date Inspected	Inspector Initials	Product Condition	Comments

Figure 5.2 Example of a rope history log. Credit: Pigeon Mountain Industries, Inc.

History Finally, well-trained professionals who work at height using ropes will carefully maintain an accurate Rope History Log for every rope. An example of a rope history log is shown in Figure 5.2. This document tracks the history of a particular rope, with space for entries that indicate the manufacturer, diameter, design, tensile strength, date of purchase, date put into service, when the rope was used, how it was used, exposure to hazardous atmospheres, and any abuse that could affect its performance or safety. The rope history log will also prove compliance with equipment inspection requirements specified by regulatory and corporate authorities.

SLINGS AND WEBBING

In addition to the main rope(s) used for rigging, other types of flexible members including rope, cordage, slings, and webbing are useful in creating anchors and auxiliary rigging. Again, it is critical to ensure that any cordage or webbing used in a life safety application is designed for that purpose. Slings and lanyards designed for utility lifting purposes should not be used for life safety rigging applications unless approved by the manufacturer. Once used for crane/lifting applications, these slings should certainly not be cross-utilized for life safety rigging or rescue applications.

Figure 5.3 Sewn anchor strap. Credit: Pigeon Mountain Industries, Inc.

Slings and webbing may be cut from a bulk spool and tied by the user to appropriate lengths and configurations, or may be used in a variety of manufactured configurations. Many life safety equipment suppliers offer both ropes and webbing that is pre-sewn into slings; some also offer pre-sewn terminations in ropes, webbing, and cordage. An example of a pre-sewn sling is found in Figure 5.3. These are particularly useful in cases where field-tied knotted terminations are prohibited by an employer or regulatory authority.

Webbing comes in two basic constructions: flat and tubular. Flat webbing tends to be thicker, stronger, and more abrasion resistant than the common alternative, tubular webbing. The increased strength of flat webbing is often preferred in harsh and abrasive environments, for use in technical rigging, and in heavy rescue situations. Since the solid flat woven construction incorporates more nylon, flat webbing weighs slightly more and can sometimes be more difficult to handle than tubular. Tubular webbing is easily identified because of the hollow center, or tube-type construction. Popular among recreational sport users, tubular webbing is lightweight, strong, and easy to handle. Tubular webbing is commonly available in a wide array of colors, and is cost efficient.

Some field personnel have a bias against tubular webbing. This has to do with manufacturing history and the progression of tubular webbing construction. Historically, tubular webbing had to be built using shuttle loom construction in order to meet "critical use" mil-spec standards. The alternative needle loom process resulted in webbing that was prone to unraveling. Because it was nearly impossible for most people to determine by appearance if tubular webbing was, in fact, built on a shuttle loom versus a needle loom, many people simply disallowed tubular webbing altogether.

These days, webbing built for life safety applications may be built using the more advanced class 1A needle loom process that incorporates a locked stitch edge, and is both efficient and reliable. However, due to historical perceptions, as well as the fact that cheap 1-inch tubular webbing made for utility purposes may not be lock stitched, many users simply forego tubular webbing. Nevertheless, tubular webbing purchased from reputable life safety suppliers and tested to the appropriate standard may be used safely. It is always best to buy webbing—and all equipment for work and/or rescue at height—from a trusted life safety product source.

Abrasion can dramatically decrease the lifespan of rope or webbing, or even destroy it completely. Usually avoidable, damage from abrasion commonly occurs when a tension-member is loaded (under tension) and dragged up, down, or sideways across an abrasive surface. Abrasion can also occur as a result of bouncy descent or ascent. The repeated load variance cause the rope to "saw" back and forth across an edge or obstruction. Use edge protection to prevent damage.

Very strong "high tech" fibers, such as para-aramids or ultra-high-molecular-weight polyethylene (UHMWPE) are also sometimes used in manufacturing slings. These should be used with great caution as they offer minimal elongation and thus are relatively unforgiving if shock-loaded. Each fiber offers its own unique set of characteristics, and should be considered specifically in light of your application and intended use.

For example, para-aramid fibers have a reported high temperature limit of up to 900+°F, but are also known to have self-destructive properties under repeated bending conditions while loaded. In contrast, UHMWPE is about 10 times stronger than steel (by weight) but one subjective consideration is that as a fiber it is so slippery that it does not hold knots well.

It is worth noting here that steel strops are also often used as anchor slings. While not completely impervious to damage, steel slings are less prone to damage from abrasion. Some steel anchor slings are further protected by plastic tubing, which shields them from sharp edges and increases the bend radius around beams. ANSI Z359 specifies that steel strops used for anchoring purposes should be terminated using two swaged fittings to help ensure security.

EDGE PROTECTION

Wherever ropes, straps, or other soft goods come into contact with edges, these should be protected. Protection is both for the protection of the rigging equipment and for the protection of the structure. Especially on buildings, façades, monuments, and other manmade structures, rope can create wear or damage where it travels over the edge. Some especially durable ropes have even been known to create grooves in limestone!

Edge protectors, also sometimes called "edge softeners," are available in several different forms. Some edge protection is intended to be used with stationary lines while other edge protection is designed to be used with moving lines. Whenever edge protection is used, it should be anchored in place either by tying it off to the structure or, if used with stationary ropes, to the rope itself.

Edge pads, consisting of a large sheet of canvas or other durable fabric, are perhaps the least expensive effective method for protecting edges (Fig. 5.4). These are best used with stationary lines, and can mold well to different surfaces.

Another type of flexible protector is a rope sleeve (Fig. 5.5). This type of protector is also made of ballistic cloth or other fabric, but is designed to wrap completely around the rope. Also intended for use with stationary ropes, it may be secured in place with a hook-and-loop or other closure. This type of protection is

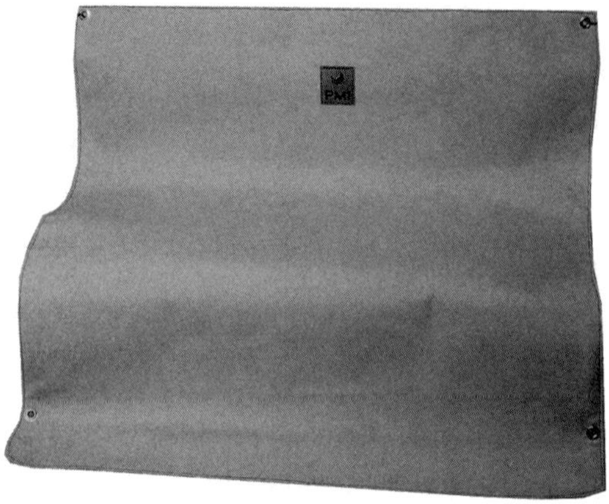

Figure 5.4 Canvas edge pad. Credit: Pigeon Mountain Industries, Inc.

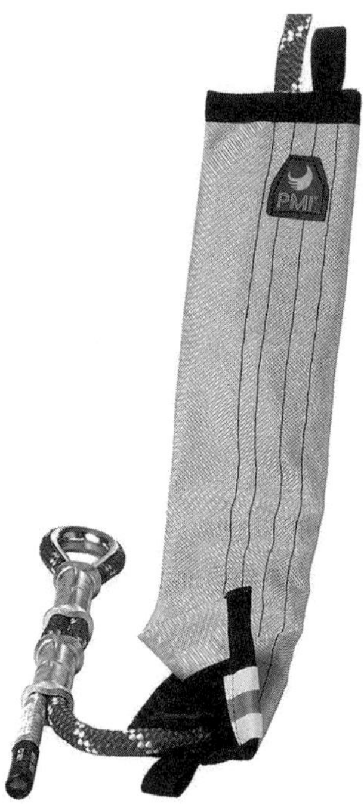

Figure 5.5 Rope sleeve. Credit: Pigeon Mountain Industries, Inc.

especially useful where a rope comes into contact with an abrasive or potentially damaging structure midline.

Plastic edge protectors are also useful (Fig. 5.6). While not as flexible as fabric, plastic does conform at least somewhat and also provides more durable protection. Plastic edge protectors may be used with either stationary or moving ropes.

Finally, edge rollers are an excellent choice for either stationary or moving ropes (Fig. 5.7). Edge rollers are available in various designs to either conform to a

Figure 5.6 Plastic edge protection. Credit: Pigeon Mountain Industries, Inc.

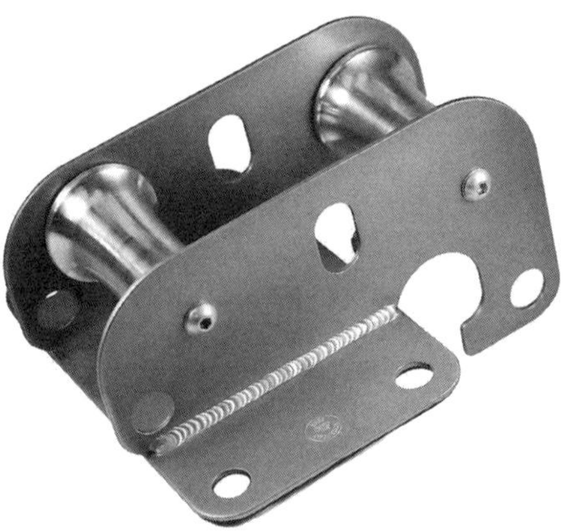

Figure 5.7 Edge roller. Credit: Pigeon Mountain Industries, Inc.

variable surface, or to effectively protect a 90° edge. Edge rollers must be affixed in place to prevent them from moving or from being dropped.

CONNECTORS

In Chapter 4 we discussed the basic idea of connectors used for personal connection to a safety or rescue system. Connectors are also commonly used for rigging systems,

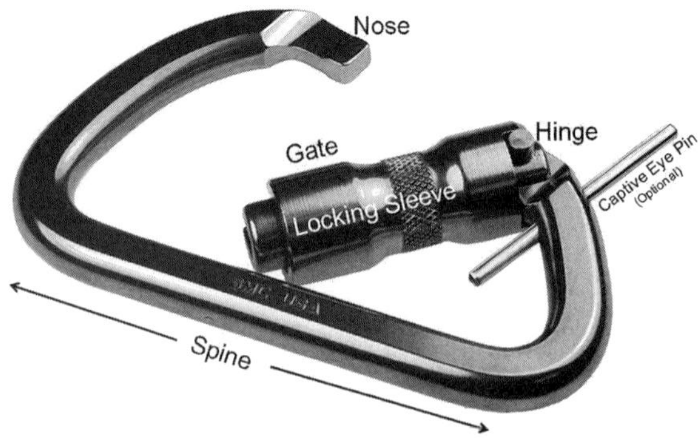

Figure 5.8 Parts of a carabiner.

and for connecting components of a system together. Connectors are available in a wide variety of shapes and sizes. These can be generally classified into the categories of carabiners and snaphooks, as described in Chapter 4.

Another type of connector, known as a screw link or a maillon rapide, is commonly used in Europe, but as these do not meet ANSI standards they are not commonly accepted for fall protection use in the United States. Experienced rescuers and rope access technicians do, however, sometimes use screw links for the distinct advantages they offer—namely, a closeable connection with greater versatility and smaller dimensions than most other connectors. If used at all, screw links should be used only with forethought and intention, and they should be screwed closed and left in place. Only screw links that are specifically designed for life safety applications should be used in these situations. Hardware store varieties of screw links do not have sufficient quality control or reliability to be used in this manner.

The strength of carabiners is generally noted in three aspects: long axis, short axis, and gate open. Parts of a carabiner are shown in Figure 5.8. The spine is the primary strength member of the carabiner. Thus, when loading a carabiner, care should be taken to align the load as much as possible with the spine. Using D-shaped carabiners (Fig. 5.9) will help to facilitate this. Pear-shaped carabiners (Fig. 5.10) are especially easy to over-stuff with more rope or webbing than can easily align along the spine.

Carabiner strength ratings are based on ideal loading, along the major axis. In real life situations, actual maximum strength of a carabiner will be influenced by two things:

1. The position of the load relative to the carabiner, and
2. Secure closure of the gate.

Some carabiners may lose as much as 75% of their strength when the gate is open. Therefore, those working at height should always use carabiners and connectors with screw-gates or self-locking methods of closure. Non-locking carabiners designed for sport climbing are not appropriate for work or rescue at height.

All connectors are subject to a relatively rare but very real phenomenon known as "dynamic rollout" (see Fig. 5.11). This can occur when one piece of hardware

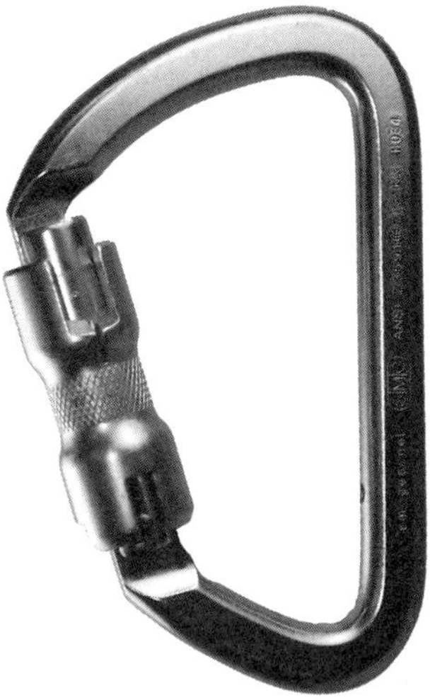

Figure 5.9 D-shaped carabiner. Credit: Donald Enos/ SMC.

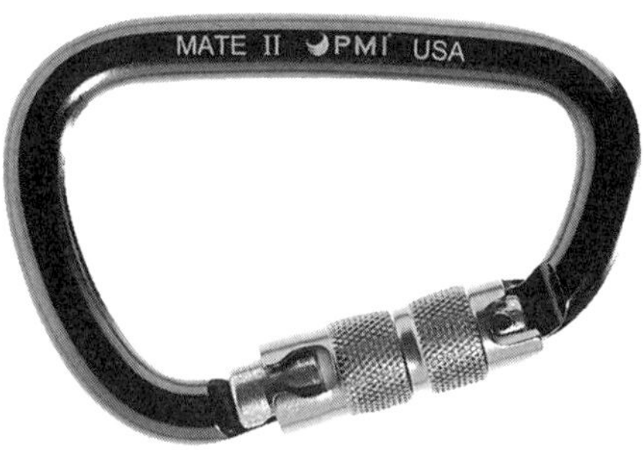

Figure 5.10 Pear-shaped carabiner. Credit: Pigeon Mountain Industries, Inc.

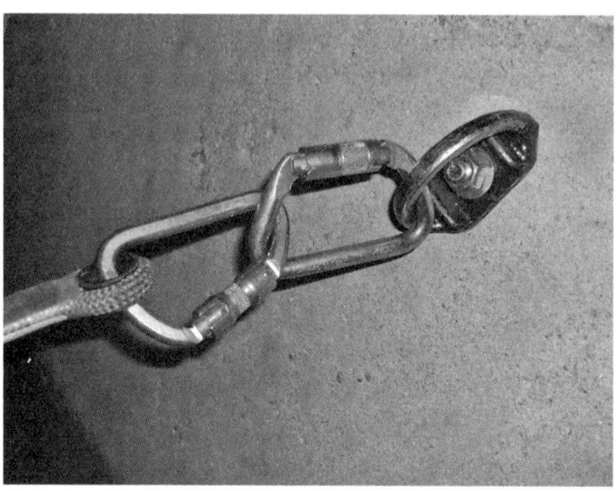

Figure 5.11 Potential dynamic rollout condition.

rotates against another, forcing the gate open. This concept was introduced in Chapter 4, but bears repeating due to the potentially catastrophic results. This is not the last time we will discuss it. Rollout is of particular concern when multiple components of rigging hardware (such as anchorage connectors or carabiners) are connected directly together, and when one of the pieces of hardware is stationary for some reason. Size is also a factor, especially when the various components are of relative size to one another to promote pressure on a carabiner gate. Such a situation might occur, for example, when a carabiner is connected to a stationary eye-bolt on a solid surface. There is no hard and fast rule to follow that will eliminate the potential for dynamic rollout. It can only be prevented by cautious and prudent rigging.

In an effort to compensate for the potential for dynamic rollout, in 2007 the ANSI Z359 fall protection standard adopted the extraordinary gate-strength requirement of 3,600 lbf. This requirement has led to development of several new connectors to meet the standard. Some of these connectors are more challenging to use than others, however, so take care in selecting which is best for your application.

Some points to consider when rigging:

1. When clipping two connectors together, avoid excessive rigidity, which can lead to rollout. If in doubt, use a soft link (described in Chapter 6) to help prevent rollout.
2. Carabiners used for clipping into an anchorage (e.g., eye bolts or shackles) should be of a design and size to allow free rotation within the anchorage without hindrance and without compromising the anchorage.
3. Verify compatibility between the size/shape relationship between any two pieces of linked hardware and how they are used to avoid dynamic rollout.

Compatibility between materials is a consideration in rigging carabiners, particularly if left attached for extended periods of time. Carabiners used to clip into steel cable, shackles or eye bolts should also be constructed of steel or other suitably hard metals.

> Carabiners are designed to be loaded directly along the spine, not along the gate. For maximum strength, the gate must be closed and, if applicable, locked. Three-way loading the major axis results in as much as 30% reduction in strength, and gate loading can be even worse.

BRAKING/DESCENT DEVICES

Braking devices for personal use in descending were briefly discussed in Chapter 4, along with connectors and other personal equipment. Braking devices may also be used for lowering a rescue load. In a lowering configuration descenders may be considered more of a system component than personal equipment.

Take the following into account when selecting a braking device for rescue:

- Magnitude of the load for which the device is designed
- Compatibility with rope and other equipment

- Whether the device is most conducive for descending or lowering
- Whether the device offers autolocking and/or panic-lock features
- Whether the amount of friction may be adjusted during use for heavier loads or greater rope weight.

Many excellent braking devices are intended only for single-person loads. Always verify that a device has sufficient friction to handle your intended load before putting it into a system.

Descenders selected for co-worker assisted rescue must be compatible with ropes and other equipment used in the system. Given the variety of devices as well as the variability of use conditions and environments, it remains to the advantage of the employer and the employee to verify that the descent device used for a given job is appropriate to that application and his or her needs.

Co-worker assisted rescue is quite different from professional rescue, and experienced technicians may utilize different types of descenders for each type of work. While professional rescuers often use free-running descenders such as the brake rack and brake tube, descenders used for co-worker assisted rescue usually offer less variability, but more fail safe operation.

A descender may be defined as an automatic or manually operated device used in conjunction with an appropriate descent line, which allows the user, either on his or her own or assisted by another person, to achieve a descent with a controllable velocity. Automatic descent devices, also known as *controlled rate descenders* (Fig. 5.12) may be further defined as a descender that features an automatic control mechanism and requires no intervention by the user to descend. These are useful for personal escape and evacuation but may be of limited usefulness for co-worker assisted rescue.

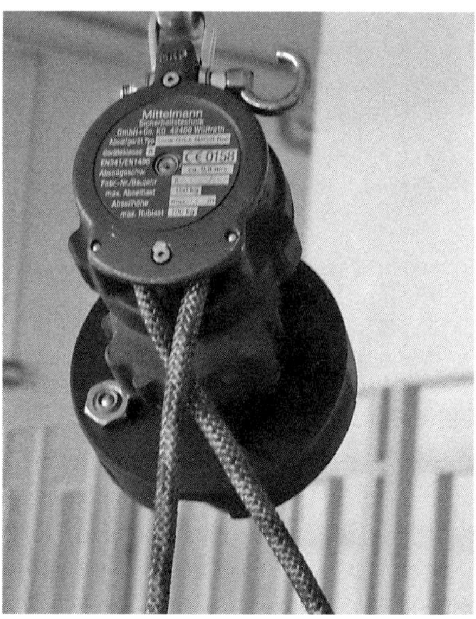

Figure 5.12 Controlled rate descenders.

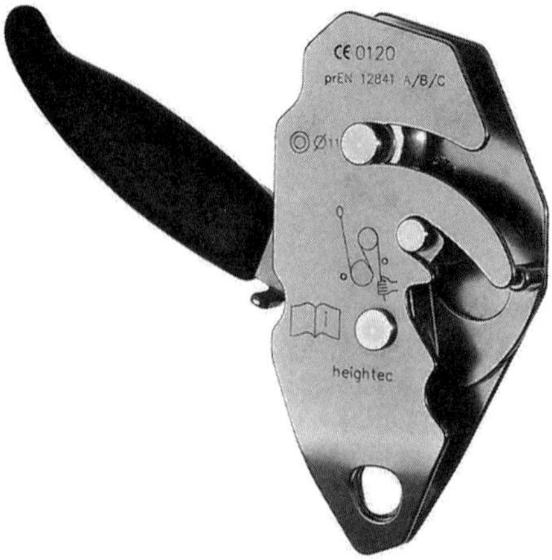

Figure 5.13 Manually controlled descender with auto-lock and panic lock. Credit: Heightec.

Manually controlled descenders (Fig. 5.13) are more appropriate for assisted rescue, as this permits more precise control and adjustment than the automatic descender. With a manually controlled descender, the operator/rescuer controls the operation of the descender and the rate of speed through some mechanical means, such as the pulling of a lever. Although even manually controlled descenders may incorporate mechanical means of operation, it is always good practice for the user to also maintain a grip on the tail end of the rope.

Some descenders incorporate an integrated descent line, while others are individual components that may be rigged onto any rope. The technician must verify that the descender he or she is using is compatible with his or her rope and with the forces, length of drop, and techniques he or she has planned.

In addition to system compatibility, three basic criteria should be considered when selecting a descender:

1. Variable Friction: When the technician is frequently changing the load he or she is carrying, or when the weight of the rope beneath him or her varies (such as on a long drop), a device with variable friction is useful. Such a device goes beyond a simple "stop/go" function to include ability to wrap, clip, or add friction components as necessary.
2. Hands-Free Locking Mechanism: A hands-free locking mechanism allows the technician to simply release his or her grasp on the rope without fear of plummeting to the ground.
3. Panic Lock Mechanism: In the event of an emergency the natural human reaction is to hang on tightly. In some descenders, however, a tight grip will actually decrease friction and cause a quick descent. A descender with a panic lock is one that stops when grasped or pulled too tightly.

A device that effectively incorporates all of these features into a single unit has yet to be developed, so priorities should be determined during the planning phase of the job so that the appropriate device may be selected.

TABLE 5.3 Descender Types as Defined by ISO 22159

Type 1: Automatically operated descending device with integrated descent line;
Type 2: Manually operated descending device with integrated descent line;
Type 3: Manually operated descending device with mechanically variable friction, hands-free locking, and panic locking features;
Type 4: Manually operated descending device with mechanically variable friction and hands-free locking features;
Type 5: Manually operated descending device with mechanically variable friction and non-automatic locking; and
Type 6: Manually operated descending device with non-mechanically variable friction and non-automatic locking.

Unless variable friction is a priority, manually controlled descenders should ideally incorporate both a panic-lock mechanism and a hands-free mechanism to guard against the potential for an uncontrolled descent. A panic lock mechanism causes the device to stop descent when engaged too fully (e.g., gripped too hard), whereas the hands-free mechanism will cause a device to stop when the operator releases control (e.g., let's go altogether).

Table 5.3 lists descender types as defined by ISO 22159, *Personal Equipment for Protection against Falls—Descending Devices.*

At a minimum, descenders should:

- Give the user suitable control over the speed of descent.
- Not cause undue shock loads to the rope when braking.
- Not cause abrasion, plucking, or stripping of the sheath under normal or expected use.
- Be of a type that cannot become accidentally detached from the rope.

While auto-locking devices are a good choice for co-worker assisted rescue applications, benefits afforded by a variable friction device such as the *brake bar rack* (Fig. 5.14) should be considered by experienced users. The rack:

- Offers greater friction, and therefore greater control, than most descenders.
- Allows the friction level to be changed after the descent has begun.
- Affords variable friction, thereby enabling the technician to safely negotiate longer drops than with most other descenders.

ANSI Z359.4, which identifies some appropriate criteria for assisted-rescue descenders, suggests that descenders should be approved for at least a capacity of 310 lb, should possess a rated strength of at least 2,700 lbf, should permit a controlled rate of descent between 1.6 ft/second and 6.6 ft/second, and should be capable of withstanding a 2 foot free fall with a 220-lb weight and still be able to function.

Descenders may be rated according to something called *descent energy.* Descent energy is a calculated rating of a device that provides information to the

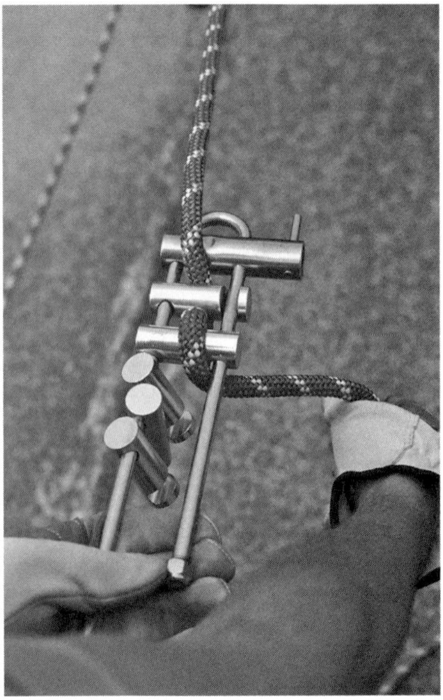

Figure 5.14 Brake bar rack. Credit: Pigeon Mountain Industries, Inc.

user regarding the weight capacity of the device, descent height, and number of descents. The formula for calculating descent energy is:

$$E = W \times H \times N$$

where

E = descent energy rating (ft/pound)

W = test weight (pounds)

H = descent height (feet)

N = number of descents.

Single-use devices should have a minimum descent energy rating of 30,000 ft/lb, while devices intended for multiple uses should be rated at a minimum E of 300,000 ft/lb.

ROPE GRABS

Personal ascenders, discussed in Chapter 4, are for ascending and personal loads only. These are intended for incremental ascent of a rope with only one person's body weight, not for hauling systems in which forces are multiplied. The toothed

Figure 5.15 Camming mechanisms of rope grabs designed for heavier loads.

cam that grips the rope, combined with the pivot action of the cam against the frame of the ascender that tightens the grip, can damage the rope if overloaded. Therefore, these should never be shock-loaded. Most personal ascenders are designed for use with 8–13 mm rope, can be easily operated with one hand, and feature a safety mechanism to prevent accidental disengagement from the rope.

Rope grabs intended for more rigorous rigging use should meet higher performance criteria. The camming mechanisms of these devices are designed with heavier loads in mind, and tend to spread the load more consistently across the rope (Fig. 5.15).

Breaking strength of a rope grab is best determined by testing it on the rope upon which it is designed to be used. At this writing ANSI Z359 specifies rope grabs for fall arrest but not for rigging. For professional rescue applications where rope grabs are used as part of rigging, NFPA 1983-12 specifies that a rope grab must hold 11 kN (2,473 lbf) without permanent damage to the device or the rope it is tested on. There is no ANSI test method for this type of rope grab.

In the absence of specific criteria for rope grabs used for co-worker assisted rescue, it is wise to choose rope grabs that meet NFPA 1983 requirements (or equivalent), for rope grabs used in rigging, such as in haul systems or to hold a system in place. Rope grabs defined by industrial fall protection standards are tested and certified for fall arrest generally in accordance with ANSI Z359. Fall arrest rope grabs should not be used in tension under normal rigging circumstances.

PULLEYS

Pulleys (sometimes called a “block” in the cranes/lifting industry) are used in rescue systems to change the direction of a rope, either to facilitate force multiplication in raising systems, or for the purpose of redirecting a load path or avoiding friction. Pulleys for rescue use should be of a design and manufacture that is intended for life safety purposes. Figure 5.16 shows parts of a pulley.

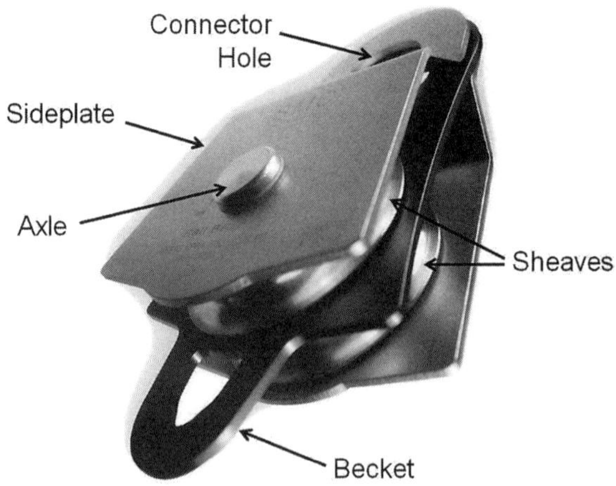

Figure 5.16 Parts of a pulley.

By using pulleys to create a mechanical advantage system, a responder can increase his or her lifting ability for the purpose of moving him- or herself, the subject, equipment, or some combination of those. These rigging concepts are explored in further detail in Chapter 8. Pulleys are designed with either fixed cheeks or swing cheeks. A swing cheek pulley is more conducive to putting on and taking off the rope midline. Some pulleys have a single sheave (Fig. 5.17), while others offer two side-by-side rope paths (Fig. 5.18).

Figure 5.17 Single pulley. Credit: Pigeon Mountain Industries, Inc.

Pulleys are designed to accommodate ropes of certain sizes, and this information is generally noted on the pulley itself. The pulley sheave is important in two dimensions: width and diameter. The width of the sheave (also called "tread width") at least in part dictates the diameter of rope that may be used with that pulley (Fig. 5.19). Using a pulley with too narrow a sheave for the rope can result in poor efficiency and possible damage to the pulley, the rope, or both. Generally a pulley with a sheave width intended for a given size of rope may be used with a smaller rope, so long as the smaller rope is not in danger of becoming jammed between the sheave and the side plate.

Figure 5.18 Double pulley. Credit: Pigeon Mountain Industries, Inc.

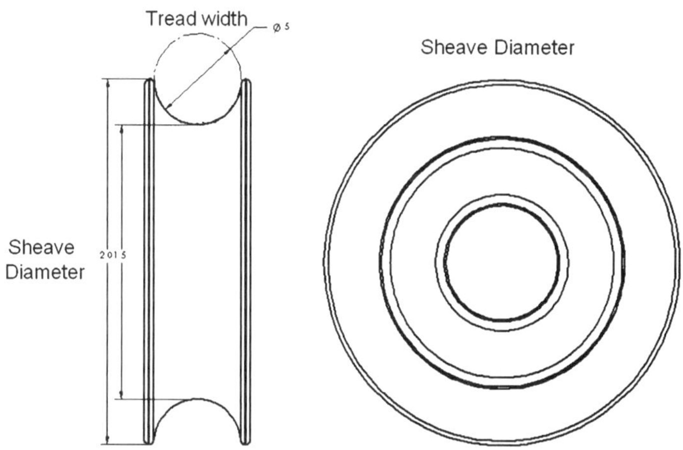

Figure 5.19 Sheave width and sheave diameter. Credit: Garin Wallace/SMC.

Sheave diameter is also an important consideration, although with modern equipment technology this dimension is less critical than it used to be. The relationship between sheave diameter and rope diameter is known as the D:d ratio. A D:d ratio of 8:1 used to be considered standard, to help reduce strength loss due to compression and elongation of the rope fibers. Modern kernmantle ropes are stronger and more forgiving, but a D:d ratio of at least 4:1 is still considered desirable.

Efficiency of the pulley itself is most dependent on the quality of bearing used in the wheel (discussed below), although sheave diameter does play a part here as well. All other things being equal, a pulley with a larger sheave will offer somewhat better efficiency than a smaller pulley.

In normal use, pulleys can easily be exposed to two times the force of the load. A firm understanding of this concept is essential in selecting a pulley of sufficient strength for the intended purpose.

A certain amount of friction is inherent in any equipment that is connected to a rope, and pulleys are no exception. In the case of pulleys, there is friction between the pulley sheave and the rope, as well as friction in the ability of the sheave to turn on its axle. This friction is referred to as pulley efficiency. Bearing type, sheave size, and side plate connections all contribute to pulley efficiency.

MULTI-FUNCTION RESCUE DEVICES

The availability of equipment with multifunction capabilities is also worth noting here. Devices that are designed to be capable of facilitating more than one task can reduce equipment load and simplify systems. For example, the Heightec Quadra, shown in Figure 5.20, can be used to lower a load, to belay a load, as a pulley, and as a progress capture device.

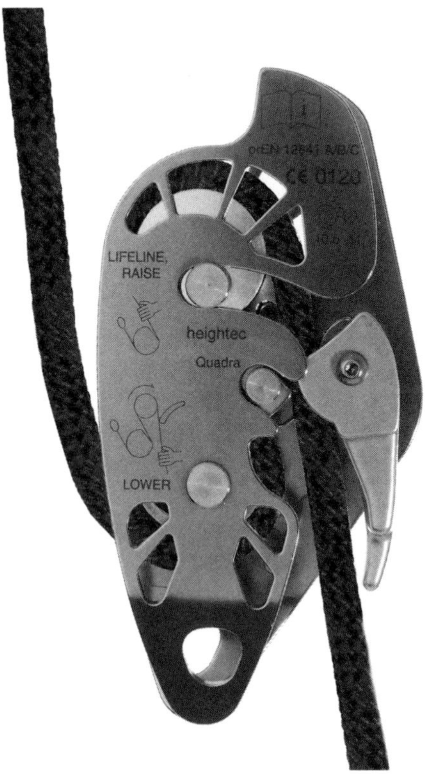

Figure 5.20 A multi-function rescue device. Credit: Heightec.

SUBJECT HOLDING DEVICE

In most cases, the subject of a co-worker assisted rescue will already be wearing a fall arrest harness, and the responder may use this harness to connect, lift, and lower the subject as necessary.

If a subject is not wearing a harness, or if the harness is not conducive to rescue, responders may need to arrange some other means of connecting him or her to the rescue system. While some texts explain how to tie an improvised harness with webbing, improvisation should not be part of a rescue *plan*, so here we will discuss other commercially available options.

ANSI Z359.4 addresses something called an "evacuation harness," which may or may not be configured like an actual harness. An evacuation harness is normally comprised of straps and material that may be secured around the body in a manner for the purpose of lifting and transporting a subject in a more or less upright position. An evacuation harness is best used with an ambulatory subject.

In its simplest form, an evacuation harness may be nothing more than a triangular shaped configuration of straps and material that can be applied upon the subject in a diaper-like manner to create a one-size-fits-most seat. In this arrangement, the evacuation harness is commonly referred to as an *emergency seat* (Fig. 5.21).

Other evacuation harnesses look more like a standard full body harness, with straps that clip around the subject to help provide security and minimize patient

Figure 5.21 Emergency seat. Credit: Pigeon Mountain Industries, Inc.

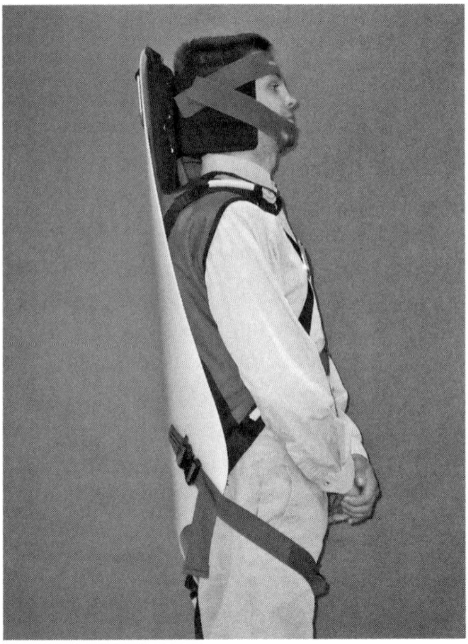

Figure 5.22 Wrap-Evac.

movement. Still other evacuation harnesses are combined with other features. For example, the Oregon Spine Splint combines a harness with immobilization features to help immobilize a subject for extrication. Similarly, the Wrap-Evac combines a lifting harness with immobilization features that harness built into a protective drag sheet for dragging along confined space and other horizontal surfaces (Fig. 5.22).

LITTERS

Subjects who are more severely injured may need to be transported in a more substantial conveyance, such as a rigid or semi-rigid full-body *litter*. Litters are available in a variety of designs and materials. Those of a rigid design, usually constructed of hard plastic, steel, or titanium (Fig. 5.23), offer more substantial support and greater protection for the subject, while semi-rigid litters (Fig. 5.24) may be easier to store and transport.

Rigid litters may be of one-piece construction, or may offer a split-apart feature for ease of storage and transport. The shape of rigid litters may also vary, ranging from a narrow 19 inches for confined space applications to a more common 22-inch width. Some litters are relatively "square" in shape, while others feature a tapered foot-end, to help shave off weight and bulk. Note that large, squared off backboards will not fit in a tapered litter. Tapered backboards are available, and may be used in this type of litter.

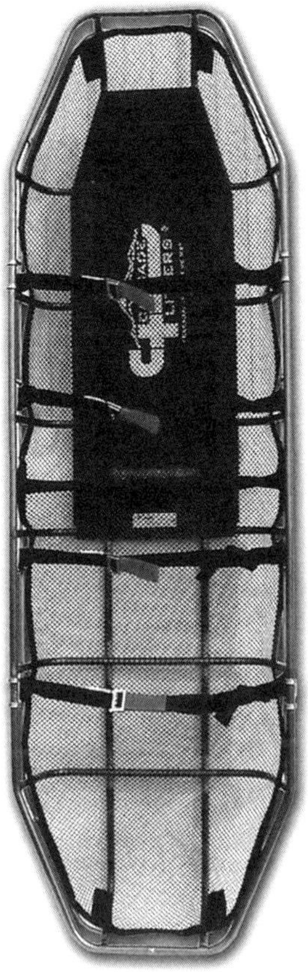

Figure 5.23 Rigid Cascade titanium litter. Credit: Pigeon Mountain Industries, Inc.

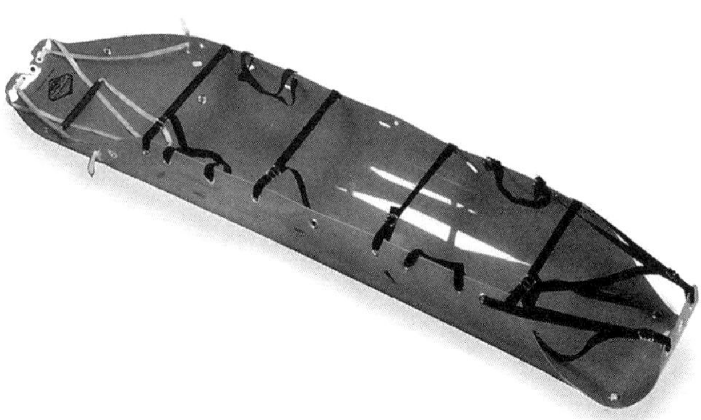

Figure 5.24 Semi-rigid Skedco litter. Credit: Pigeon Mountain Industries, Inc.

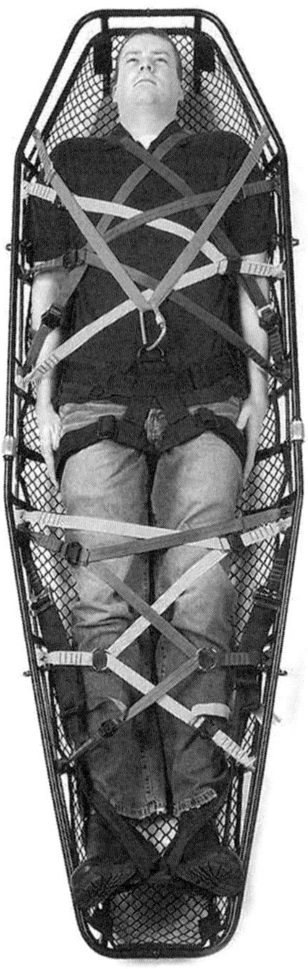

Figure 5.25 FAST patient restraint. Credit: Pigeon Mountain Industries, Inc.

Regardless of design, litters should be equipped with some effective means of securing the subject in place during transport. While this can be accomplished with a long piece of webbing and a series of knots, commercially available alternatives such as the FAST (Fig. 5.25) may be more secure and easier to use. Where a litter is used for vertical lifts or lowers, the method of securing the subject is especially important, and rescuers should ensure that the subject cannot be ejected from the litter even if it should become inverted or tilted in any direction.

There is no one type of subject holding device that is ideally suited for all applications. The employer should select the type, size, shape, and material of harness alternatives based on anticipated need given the type of incident(s) that are reasonably foreseeable.

A litter used for high angle lifting or lowering should also have some means of being securely attached to the rescue system in a manner that will provide stability and balance, and that cannot be inadvertently released while under load. Commonly referred to as a litter spider, some of these are designed to support the litter in a horizontal configuration (Fig. 5.26), while others suspend the litter vertically.

Figure 5.26 Litter spider provides horizontal litter position for high angle rescue. Credit: Pigeon Mountain Industries, Inc.

PRE-ENGINEERED RESCUE SYSTEMS

A pre-engineered rescue system is one that has been designed by the manufacturer or other Qualified Person, combining multiple components into an approved system to work effectively together for a given rescue purpose.

The simplest type of pre-engineered system for a layperson rescuer to use is a retrieval system. A retrieval system (Fig. 5.27) is designed to retrieve a worker from a hazardous condition by a simple action on the part of a co-worker, such as turning a crank, without entry into the hazardous area by the co-worker. Retrieval systems perform best where there is a single entry/exit path for the worker, where egress is a straight-line lift or lower, and where the worker is monitored at all times by a co-worker who is prepared to initiate retrieval immediately. These types of systems are most often used successfully in confined space operations.

The basic idea behind a retrieval system is that the co-worker performing the retrieval needs only a minimal amount of training. The worker may use a ladder, rope descent system, suspended scaffold, or any of several methods of access, but the retrieval system should be separate from the method of entry. The worker should be attached to a retrieval system prior to entering the high angle environment. In the event of an accident, or unresponsiveness on the part of the worker, he may be pulled by a co-worker to safety by his retrieval system. Retrieval systems can be an excellent choice for speedy, simple retrieval operations.

The foundational part of a retrieval system is the anchorage with an attachment point that is suitable for lifting/lowering as necessary. This may be a structural

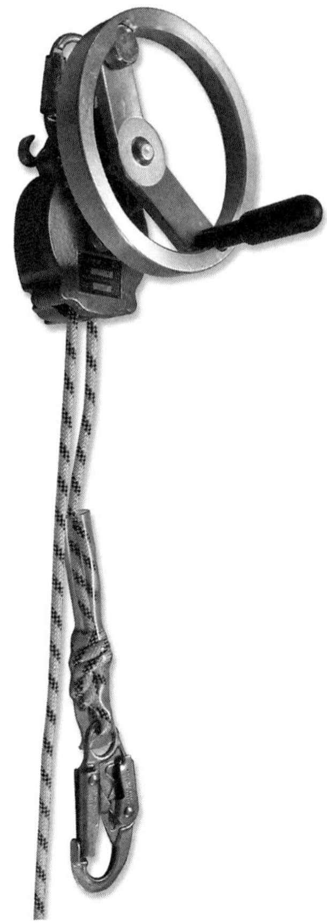

Figure 5.27 DeRope retrieval system. Credit: Tractel, Inc.

anchor at the workplace, or a portable device such as a tripod, quadpod, or davit system. The retrieval system should be anchored separately from any fall protection point, if one is used, although both the fall protection system and the retrieval system may be anchored to the same anchorage. You will find further discussion of anchorages, anchorage points, and related information in Chapters 8 and 9.

The actual retrieval action is usually performed using a hoisting mechanism with mechanical advantage. Regardless of the type of mechanism used, a retrieval system should be capable of both raising and lowering, and should have a fail-safe mechanism to prevent loss of control by the person performing the retrieval.

Co-worker assisted rescue should, if possible, be performed without placing the rescuer at risk. To this end, using a pre-engineered system that allows the rescuer to connect to the subject remotely (Fig. 5.28), without entering the hazard, is generally preferred over any method that requires the rescuer to expose him- or herself to a hazard—whether a vertical environment, confined space, or other exposure.

Specific methods for achieving a remote connection and retrieval in the fall protection environment will be addressed in Chapter 8.

Figure 5.28 Remote connection. Credit: Vertical Rescue Solutions/PMI.

CARE, INSPECTION, AND RETIREMENT OF RESCUE GEAR

Retirement of life safety rope and related equipment is a serious proposition—one upon which your life (and the lives of others) may depend. Most manufacturers offer at least some basic guidance for inspection, but in fact inspection of life safety equipment is a highly subjective process, and proficiency can be gained only through a combination of training and experience.

Inspection of soft goods, such as rope, webbing, or sewn equipment, is an especially subjective practice, and pass/fail criteria may vary depending on the use conditions, environment, and desired future performance. Each component of equipment should be closely inspected, and this process requires attention to unique details. While this text provides some general concepts for inspection, anyone who may be expected to inspect equipment should be properly trained to do so by a Competent Person Trainer.

Use the following steps as a guide when inspecting hardware such as connectors, descenders, and other metal goods:

1. First, simply look at the device. Ensure that it is not corroded, cracked, or deformed. There should be no missing or damaged parts.

2. Look closely at wear points where the device comes into contact with other components during use. Ensure that contact points are not excessively worn, and that wear is not apparent on parts of the device that do not normally come into contact with other equipment.
3. Feel the device in your hands, verifying what your visual check has noted. Sometimes irregularities can be felt more easily than they can be seen.
4. Verify that all pins and other connections are properly seated and intact.
5. Next, check any moving parts to verify that gates, levers, or other actions perform as designed, without sticking or hesitation, and that they are not loose.
6. Finally, connect the device to a rope or whatever equipment that component is designed to be used with, and verify that it functions properly.

Software can be a little more challenging to inspect. The Cordage Institute offers guidance on performing visual and tactile rope inspection through document number CI 2005 Guidelines on Inspection and Retirement of Life Safety Rope. This guide provides an excellent baseline for inspecting soft goods through visual and tactile methods. General guidelines for inspecting soft goods include:

1. Inspect the condition of the entire length of the fiber component (webbing or rope); ensure that it is not discolored, torn, abraded, or excessively worn.
2. If the equipment is stitched or otherwise terminated, check the condition of the termination. These should be intact, with no torn or cut stitching, and free from excessive wear. Pay special attention to load-bearing stitching.
3. If the piece of equipment you are inspecting incorporates other parts (buckles, abrasion guards, etc.) inspect contact points closely. Manipulate the parts as necessary to visually inspect every centimeter of the fiber parts.
4. Next, inspect other critical parts closely. If there are buckles, connectors, or other attachments, these should be free from damage or excessive wear, and should function properly.
5. Also check protective or convenience parts to ensure that they are intact.
6. Finally, verify the function of the equipment, ensuring that it operates as originally intended.

Maintaining a use log where information about each piece of equipment can be documented is a good way to keep detailed information about other conditions that may affect the performance of the rope: exposure to chemicals, impact forces, high loads, age, and so on.

One easy determining factor for discarding a piece of equipment is if it has been subjected to high forces or uncontrolled loading. If any item has experienced a significant load, it is best to simply assume that that piece of gear is no longer viable. This same rule should also hold true for any other equipment that might have been in the same system at the time. The best advice to follow in inspecting any life safety equipment is: "If in doubt, throw it out." Any item of life safety equipment that does not look right, does not feel right, or that you simply do not feel right about, should be retired—even if it passes "inspection" with flying colors.

SUMMARY

Rescue equipment includes any equipment that is used in a raising, lowering, or other system to transport a subject, especially in a high angle environment. Criteria for selecting rescue equipment should be specific to its intended use and application. While certain equipment standards may apply, and can provide limited guidance, it is the use criteria that should drive the decisions.

In general, a conservative approach is recommended when it comes to selection, inspection, and retirement. Equipment must not only be appropriate at a component level, but all of the components together must be compatible. Selecting pre-engineered systems is a good way to help ensure compatibility of components.

Equipment should be maintained according to manufacturer's instructions, and should be discarded if there is any doubt as to its serviceability.

POST-CHAPTER QUIZ

Chapter 5

1. Equipment requirements may vary based on:

   a) The worksite

   b) Potential incident type

   c) Environment

   d) All of the above

2. True or False: Any item that is approved for lifting (such as with a crane) can be assumed to be appropriate for life safety applications.

3. Minimum breaking strength of life safety equipment components is generally specified as

   a) The safe working load that is stamped on the equipment

   b) Three standard deviations below the mean breaking strength of a minimum number of samples of that component when tested

   c) The average breaking strength of a minimum number of samples of that component when tested in accordance with industry standards

   d) The lowest breaking strength of a minimum of five samples

4. True or False: A rope with a stiffer "hand" will usually offer greater abrasion resistance.

5. Dynamic rollout occurs in a hardware connector when

   a) A person uses a round connector

   b) A connector is not used

   c) One piece of hardware rotates against another, forcing the gate open

   d) None of the above

6. The ideal way to load a carabiner is
   a) Along the short axis
   b) Across the gate
   c) Against the nose
   d) Along the spine
7. Manually controlled descenders should ideally incorporate
   a) A panic lock mechanism
   b) A hands-free mechanism
   c) Both a panic lock and a hands-free mechanism
   d) An integrated pulley
8. Things that contribute to the efficiency of a pulley include (check all that apply)
   a) Bearing type
   b) Sheave size
   c) Side plate connections
   d) Color
9. The basic idea behind a retrieval system is that
   a) The co-worker performing the retrieval needs only a minimal amount of training
   b) It should be used only for a deceased subject
   c) It is only appropriate for use in confined space operations
   d) All of the above
10. The best advice to follow in inspecting any life safety equipment is
   a) If it has ever been used, discard it
   b) If in doubt, throw it out
   c) If it looks okay, it is probably fine
   d) If it has not been used in over a year, throw it out

ANSWER KEY

1. d
2. False
3. b
4. True
5. c
6. d
7. c
8. a, b, c
9. a
10. b

CHAPTER 6

PERSONAL SKILLS FOR RESCUERS

By the time you are finished with this chapter, you should understand:

- The importance of rescuer safety.
- How to form a few basic knots, and why this knowledge is important.
- Basic anchoring skills.
- How to descend a rope.
- The basics of free climbing and ascending a rope.
- How to rig a backup safety system.
- The basic requirements of medical care of a subject.

In the moment of a true emergency it is the most basic, ingrained reactions that take over. A person who has rehearsed and prepared in advance is much more likely to respond proactively and with better results than one to whom the crisis comes as a complete surprise. While administrative aspects of advance planning are covered in Chapters 1–3, and equipment is discussed in Chapters 4 and 5, preparation must also include hands-on practice to help ensure appropriate capabilities on the part of the individuals on the worksite. The next few chapters will provide guidance in practical skills for responders.

Any person who is working at height, or working near others who are working at height, has the potential to become a rescuer. Imagine that as you are reading this chapter you suddenly hear a loud bang outside, followed by screaming. You leap from your chair and run outside, where you see a lopsided scaffold dangling from the building next door. Two window cleaners are dangling from their fall arrest systems, suspended by the dorsal attachments of their harnesses. Their swing stage sways precariously from a single cable several feet below. Palms sweating, you dial the local emergency services number and burble out something that you hope is close to the actual address where you are. The dispatcher tells you that the fire department will arrive in 10 minutes.

Outside on the street below a crowd is gathering. One of the dangling window cleaners hangs limp, blood dripping from his head. The other appears uninjured, but is flailing around and cannot seem to rescue himself. His continued screaming is

Falls from Height: A Guide to Rescue Planning, First Edition. Loui McCurley.

some comfort (it assures you that he has an airway) but within a few minutes he begins losing energy, grows quieter, and it becomes apparent that he is struggling to breathe as he awaits rescue.

Now imagine that the person dangling from the fall protection system is someone you know. Imagine that he is a co-worker with whom you have worked for several years, perhaps a good friend. You may have been at his house for dinner just last weekend, your kids playing together outside. What pressure you would feel to find some way to help!

RESCUER SAFETY

Knowing what to do in an emergency begins with knowing what *not* to do. Things to not do in an emergency include panic, run, enter a hazardous environment, or perform a skill that is beyond your level of ability and experience. Rescuer safety begins with proper prioritization. As a rescuer, one's own safety comes first, the safety of co-rescuers comes second, and that of the subject is a distant third. This fundamental concept is foundational, and might be called the "rescuer's safety maxim."

Accepting responsibility for the welfare or rescue of another person in a hazardous environment is a weighty prospect and should not be undertaken lightly nor without proper preparation. A would-be rescuer must be trained and experienced in the environment and under the conditions at the site of the rescue. This is one reason that co-worker assisted rescue can be so effective as an immediate response plan: co-workers within a workplace are usually more experienced, more familiar with, and more prepared for specific worksite conditions and hazards than are external responders.

In addition to having a firm understanding and ability to protect him- or herself from a fall, and whatever additional skills are necessary to work in the environment where he is engaged, any person working at height should be sufficiently prepared to recover in the event that he or she falls into the safety system. A person working at height should also be capable of responding to an incident where a co-worker falls into his or her safety system, in case the co-worker cannot self-rescue. A few basic skills will go far in keeping both the fallen worker and the would-be rescuer safe.

The most important things for a worker to do in the event of an emergency are:

1. Recognize and avoid immediate dangers.
2. Secure and control the site to prevent further incident/injury.
3. Activate the appropriate emergency response protocols.
4. Communicate details effectively to the person(s) in charge.

To accomplish these basic needs, a worker must be trained to recognize potential hazards associated with a workplace incident. While this text focuses specifically on response to fall incidents, other hazards may exist in combination with the fall. At a minimum, workers must be trained and capable of working efficiently and maintaining their own safety in any environment where they might intend to assist another. Some of the potential hazards that workers should be trained to recognize include environmental conditions such as hazardous materials, weather, atmospheric contaminants, and fire; equipment hazards including machinery and electrical risks; working surface hazards such as surface angle, elevation, and slip conditions; and potential disruption to infrastructure, including availability of power and communications.

Perhaps more obvious are the hazards that are specific to the incident itself. Simply the act of working in the vertical plane creates a potential "fall hazard zone," comprised of the work location as well as areas below and nearby where falling equipment or debris could hurt someone. Also important is the "access zone," or the area from which a worker enters/exits the hazardous/vertical plane. Both these zones should be given special protection when an incident occurs to help protect the subject, would-be rescuers, and bystanders.

Perhaps the least predictable aspects of a fall emergency are the people involved. The actions and effects of people, including those involved in the incident, co-workers, responders, bystanders, and those otherwise unrelated to the incident, can dramatically affect the outcome of an incident. Remaining calm is of utmost importance, and proper preplanning and training can contribute positively to appropriate response and a good outcome.

Familiarity with workplace-specific hazards will help the rescuer to recognize whether these things might have contributed to the incident, and to protect him- or herself while assisting others. Evaluating such potential hazards involves knowing what to look for and asking appropriate questions. Good up-front information will help the reporting person to secure the correct resources for the incident, and to keep him- or herself and others safe in the face of the inevitable sense of urgency present on an accident site.

ROPES AND TERMINATIONS

At least basic rope management and knot tying skills are essential for anyone who is using ropes for work at height and/or who might be expected to perform co-worker assisted rescue.

Wherever possible, *manufactured terminations* are the preferred choice for industrial fall protection and co-worker assisted rescue. Whereas the laid and braided ropes once so common in life safety applications can be spliced, modern kernmantle life safety ropes cannot. Instead, they are factory terminated with heavy metal swages, or they are sewn at the factory (Fig. 6.1), using special jigs and stitching techniques.

Because factory-sewn and -swaged terminations are produced by the manufacturer, they offer consistency and reproducibility, and they avoid the opportunity for the user to incorrectly form the termination. However, manufactured terminations do pose limitations. They do not afford the worker the opportunity to easily adjust the location of the termination for the purpose of fine-tuning his or her system; they cause the rope to be loaded repeatedly in the same location, thereby increasing point-stress; and they limit the variable functionality of the system.

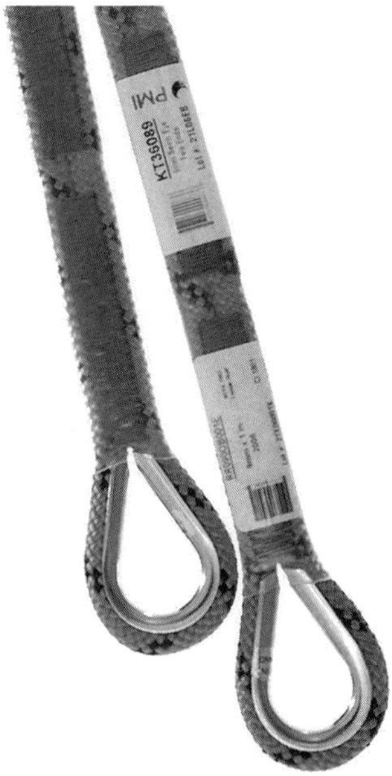

Figure 6.1 Sewn termination. Credit: Pigeon Mountain Industries, Inc.

Even if the pre-engineered rescue system that a workplace response group intends to use features manufactured terminations, co-worker assisted rescue responders should be capable of tying at least one of each of the following types of knots:

1. To form an end-loop around an object.
2. To create a midline loop in a length of rope.
3. To join two ropes.

The capability to tie these basic knots as needed will help prevent a would-be rescuer from being hampered by the inability to adjust his or her system, or by a compromised termination, during a rescue.

Only knots that are recognized industry best practice should be used for life safety purposes. Although all knots or bends in a rope will affect the strength and performance of that rope to some degree, some knots have a greater affect than others. When determining which knot to use for a given purpose, the following considerations apply:

1. Ability of the user to correctly tie the knot.
2. Appropriateness of the selected knot to the intended purpose.
3. How the knot will interface with the system (e.g., size, location, etc.).
4. Efficiency (strength) of the knotted rope.

"Knot efficiency" refers to how much of the host rope's strength is retained in the knotted portion of the rope as compared with the unknotted portion. For example,

TABLE 6.1 Strength Efficiencies of Some Knots Commonly Used in Life Safety Applications

Purpose	Name of Knot	Efficiency
Loop in the end of a rope	Bowline	63%
Loop in the end of a rope	Figure eight	73%
Midline loop	Butterfly	72%
Connecting two ropes	Double fisherman's	78%

if the breaking strength of a rope is rated at 6,000 lbs (100%), a knot that is 80% efficient can be said to reduce that strength by 20%, to approximately 4,800 lbs, at the location of the knot. The Cordage Institute has published guidelines regarding knots in the form of a technical report (Table 6.1).

Many textbooks exist on the subject of knot-tying, and the avid rescue technician will find it easy to delve into a detailed study on knots. For practical reasons, only a few of the knots most commonly used in life safety applications will be addressed here. The ability to tie a great variety of knots is much less critical than the ability to tie a few good knots very well. To ensure proper understanding of knots, hands-on training is required under the tutelage of a competent instructor.

Knot Terminology

A good understanding of some basic terminology is essential when learning to tie knots. The rope itself may be referred to as a rope, or as a *line*. A line is a rope in use. The end of a rope that is being used to tie the knot or otherwise perform work is called the *working end* while the end not presently in use may be referred to as the *standing end*, as illustrated in Figure 6.2.

The *tail* of the rope is the bit left over after you have tied a knot. A *bight* is a section of rope bent into a U shape, as in Figure 6.3. This is the starting point for many knots. A *loop* goes one step further than a bight, and is created when one end of the rope crosses over the other, forming an X and creating a closed loop (Fig. 6.4). An *overhand* knot (Fig. 6.5) is created from a loop, by passing the working end of the rope through the loop to the other side and then pulling down snugly. A *barrel* knot (Fig. 6.6) is created when the working end of a rope is passed twice around itself, then the end is passed back through the loop in the rope.

In addition to creating the foundation for many other knots, either the overhand or the barrel knot may be used as a *stopper knot* in the end of a rope, as shown in Figure 6.7. The stopper knot helps prevent someone descending off the end of a rope. Either of these knots may also be used as a *safety knot* in conjunction with other knots.

Another knot that is sometimes used as a stopper knot is the *figure 8 bend*, which is formed by taking the working end of the rope twice around the standing part, creating an 8 shape, before passing the working end through the loop (Fig. 6.8).

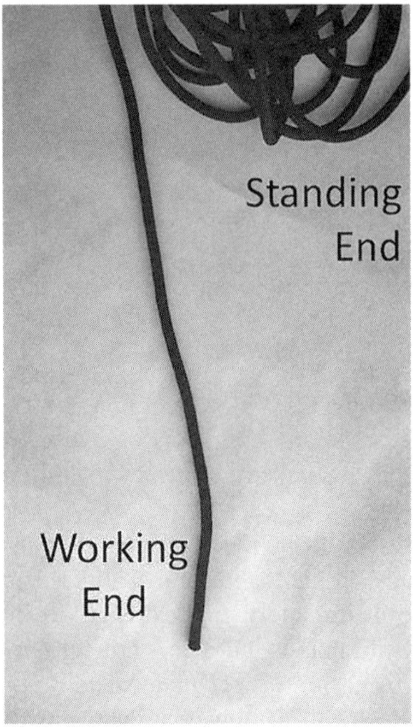

Figure 6.2 Standing end and working end of a rope.

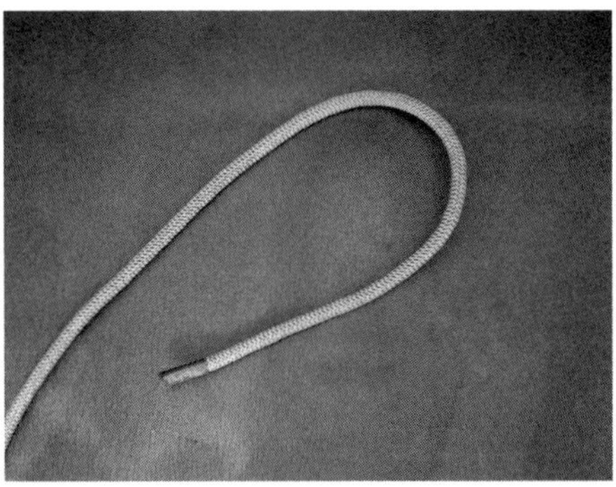

Figure 6.3 Bight.

Figure 6.4 Loop in a rope.

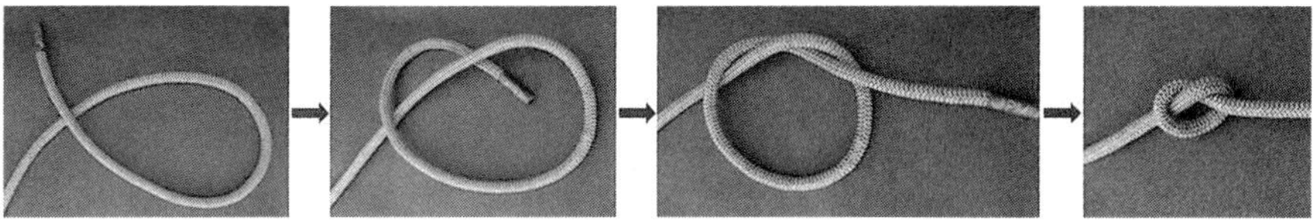

Figure 6.5 Overhand knot.

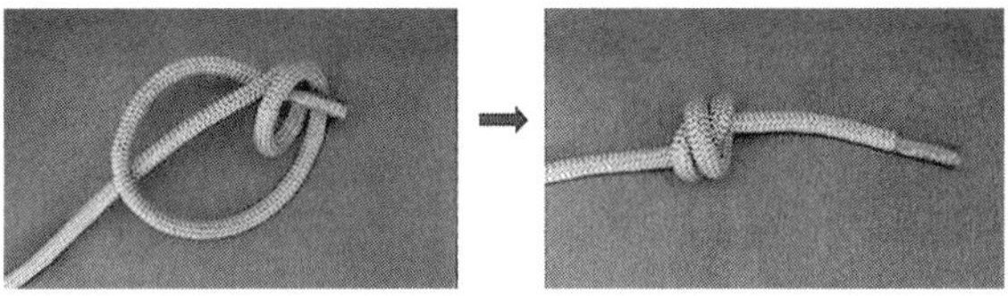

Figure 6.6 Barrel knot.

Figure 6.7 Stopper knot.

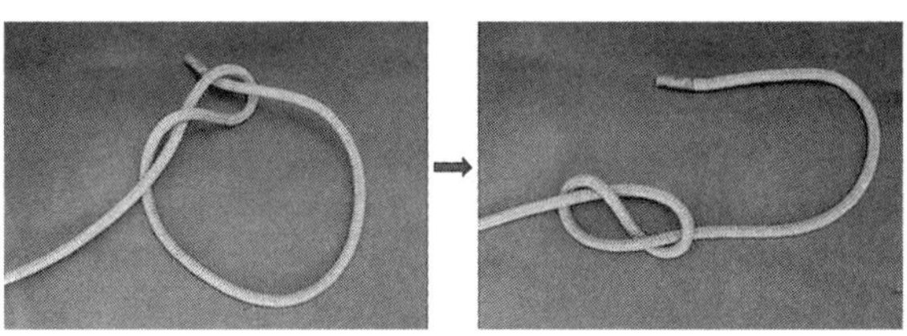

Figure 6.8 Figure 8 bend.

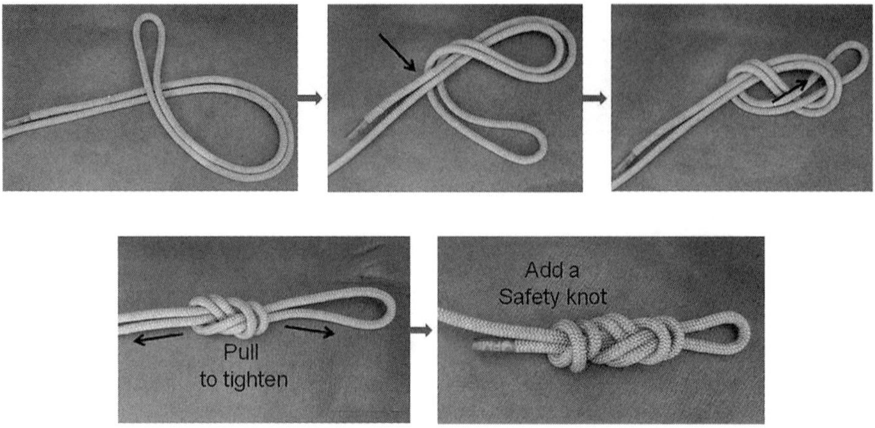

Figure 6.9 Figure 8 on a bight.

Figure 6.10 Figure 8 on a bight with a safety, connector attached.

End Knots

The figure 8 bend described above can be adapted to create an end loop in a rope so that it may be connected to something; in this case it is called a *figure 8 on a bight*. This end knot is created by simply tying the figure 8 bend described on the preceding page into a bight of rope, as shown in Figure 6.9.

Be sure to leave sufficient tail in the rope after it exits the knot. At least a hand's width is recommended. For extra security, a safety knot may be added to an end knot simply by tying an overhand in the tail of the rope as it exits the knot (Fig. 6.10). A safety knot prevents the tail of the rope from being pulled backward through the knot (thereby untying it) under tension, provides backup in the unlikely event that the knot is tied incorrectly, and helps keep the tail of the rope out of the way.

A connector may be attached to the figure 8 on a bight so that it may be used to anchor a line, to reposition a line, to connect a rope to a subject or a rescuer, or for a host of other reasons.

Also from the figure 8 family of knots, a figure 8 retrace may be used to connect a loop of rope directly to a structure, such as an anchorage. The figure 8 retrace is easy to tie, is easy for a safety supervisor to examine, and is very efficient

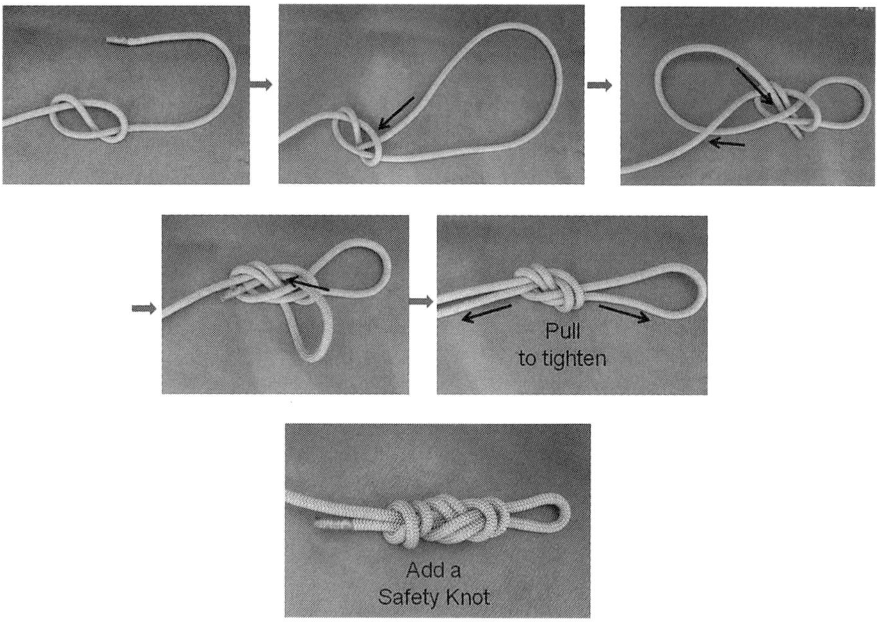

Figure 6.11 Step-by-step figure 8 retrace.

in terms of retaining strength. To tie a figure 8 retrace, follow these steps (shown in Fig. 6.11):

1. Tie a simple figure 8 bend approximately three feet from the end of the rope (further if the object one will be tying around is very large).
2. Pass the working end of the rope around the object to which it is being tied.
3. Starting where the working end of the rope exited the figure 8 bend, use the working end to "retrace" the path of the 8, ending with the tail of rope coming out alongside the mainline.
4. For maximum security, finish the knot with a safety overhand that goes around the mainline.

According to test results published by the Cordage Institute, a figure 8 can retain as much as 80% efficiency when used to form a loop in the end of a rope.

Midline Knots

Midline knots require a different approach, because the working end of the rope is generally not available to be used to help tie the knot. Midline knots can be useful for re-directing a rope, for connecting multiple people to a rope, or for isolating a damaged segment of rope. One consideration for use of midline knots is to ensure that the direction of pull on the load line is appropriate for the knot selected.

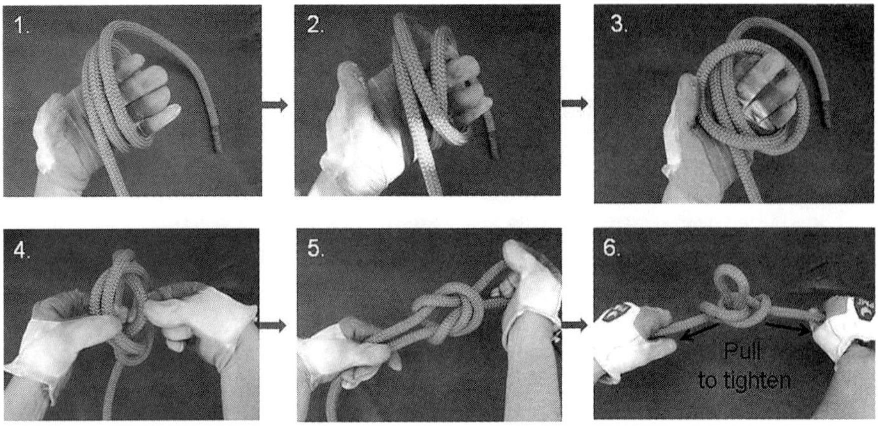

Figure 6.12 Step-by-step butterfly.

Figure 6.13 Clip a connector into the midline loop for security.

One excellent choice for use as a midline knot is the butterfly. One way to tie a butterfly knot is shown in Figure 6.12. To tie a butterfly knot:

1. Wrap the rope three times around your hand.
2. Pass the line closest to your fingers over the line in the center.
3. Pass the line that is now closest to your fingers over *both* of the other lines as they now lie.
4. Tuck the line now nearest your thumb *under* and through the other two lines.
5. Tighten.

Safety knots are generally not used with a midline knot because there is no "tail" in which to tie one. In any case, the purpose of a midline knot being to create a loop in the middle of the rope, the likelihood of either end of the rope being pulled back through the rope is minimal. Of greater concern is ensuring that the tension on the two ends of the rope does not pull the loop out of the rope. Simply clip a connector into the midline loop as shown in Figure 6.13 to provide greater security against this happening.

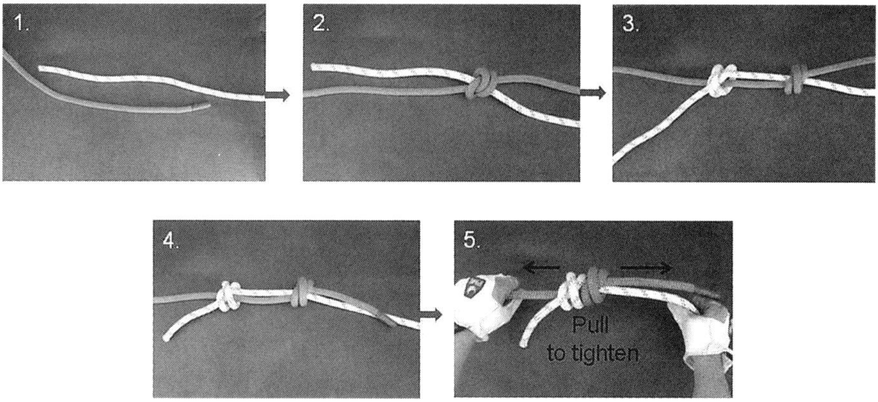

Figure 6.14 Double fisherman's bend.

Joining Two Ropes

Good planning reduces the probability of ever needing to join two ropes together, but sometimes it is necessary. In the event this does become a necessity there is really no substitute for an appropriate knot. While it is possible to link two manufactured terminations with a connector, this introduces extra components and extra bulk, making the system less streamlined and potentially introducing additional hazards.

The most secure method for achieving a low-profile, high strength connection is the double fisherman's bend. To tie this bend, follow these steps (pictured in Fig. 6.14):

1. Align the two ropes next to one another, with the ends pointed in opposite directions.
2. Wrap the first rope around the second, enclosing it with a barrel knot.
3. Now tie the second rope around the first with another barrel knot, this time in the opposite direction.
4. Snug each knot down.
5. Pull the two ropes in opposite directions so that the two knots nest against one another.

One disadvantage of the double fisherman's bend is that it can be very difficult to untie after it has been loaded heavily. It also is not the most secure method for joining two ropes of very different diameters. For joining ropes of different diameters, the double sheet bend, shown in Figure 6.15, is a better choice. To tie the double sheet bend:

1. Make a bight (U-shaped bend) in the thicker of the two ropes, aligning the two strands of the rope (standing end and working end) next to one another.
2. Pass the working end of the second rope up from underneath the first rope and through the U-shaped bend.

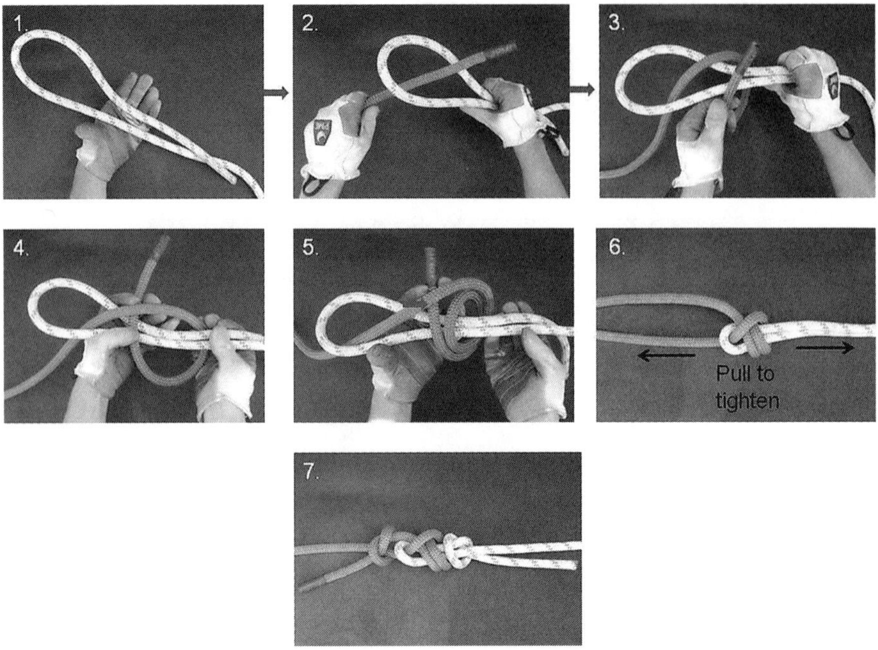

Figure 6.15 Double sheet bend.

3. Continue the end of the second rope around, behind the aligned strands, and back to the front.
4. Pass the end of the second rope under itself at the point where it first exited the U bend in the first rope.
5. Wrap the second rope once again around, behind the aligned strands and back to the front.
6. Pass the end of the second rope under itself again at the point where it first exited the U bend in the first rope.
7. Tighten the knot and use overhand or double overhand knots to secure the ends of both ropes.

ANCHORAGES

The first step in creating any system for safety or rescue at height is to establish an appropriate anchorage system. This begins with selection of an adequate anchor point. Anchorage systems, including strength and requirements, are further discussed in Chapter 7; this chapter will focus on anchoring concepts and practical application of anchor points.

Only anchor points that have been identified and approved by a Competent or Qualified Person should be used for that purpose. A good rescue anchorage is strong, secure, free of sharp or abrasive surfaces, and is in line with the desired *fall line* (see Chapter 7 for more discussion of fall line).

As the foundation upon which the entire system relies, a rescue anchorage should be unquestionably reliable, be positioned such that the rescuer can maintain his or her work position without difficulty, and be placed so that the rescuer(s) can connect to or disconnect from the rope system in an area where there is no risk of a fall from a height.

Anchorages should be positioned above, in line with, and out of the way of the point of descent. Anticipate strength, direction of loading, potential abrasion, and any other factors that may affect anchor performance.

Anchorages for rescue systems must be carefully constructed to ensure capability of withstanding whatever loads might be applied. While the strength necessary is driven by a combination of regulatory requirements, need, and availability of options, consider that a good practice to follow for co-worker assisted rescue is that which is found in ANSI Z359.4. This standard recommends using an anchorage of at least 3,000 lb (13.3 kN) strength (or, for engineered anchorages, five times the applied load) per person attached for co-worker assisted rescue. During the rescue planning phase and at the time of rigging keep in mind that anchorages used for rescue may be exposed to a two-person load and/or additional directional loading. Further, if the rescue operation could expose the rescuer to a fall hazard, an anchorage suitable for fall arrest should be selected.

In most cases, anchorages designed and intended for fall arrest are usually rated in the 5,000 lb range, which is generally also quite sufficient for co-worker assisted rescue use. Properly designed and installed anchorages should include posted information regarding intended use, strength, direction of approved loading, and so on. Such guidelines should always be noted and adhered to.

Connecting to the anchorage can be achieved in any number of ways. Engineered anchorages that have been designed and installed in anticipation of fall protection or rescue may offer an eye bolt, D-ring or O-ring connection point, such as that shown in Figure 6.16.

While installed anchorage connectors may appear easy to use, similar appearing components may be installed on equipment as lifting points, and may not be intended for life safety purposes, while others may be rated only for restraint. Unless you specifically know that a bolt or anchor ring is intended for fall protection or rescue use, do not rely on it. When using this type of anchorage connection take care to avoid rollout conditions, which are addressed in greater detail in Chapter 8.

In practical application, especially applications pertaining to rescue, eye-bolts or other permanently installed anchorage connectors are not always pre-engineered exactly where they are needed. If the anchorage is a structural member, such as an I-beam or structural column, the best means of connection may be to wrap it using a protected steel cable, fiber rope, or webbing sling (Fig. 6.17).

A sling of any of these materials may be wrapped around a suitable anchorage and both ends clipped together with a carabiner or other connector. The sling may be wrapped once, twice, or multiple times around an anchor point, but always ensure that the Y-angle formed at the tangent end of the sling is 90° or less and that the connector is locked (Fig. 6.18).

Figure 6.16 D-ring anchor point.

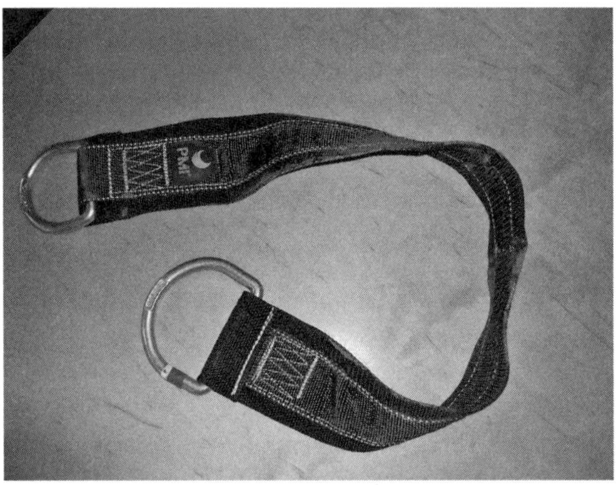

Figure 6.17 Webbing sling anchorage connector.

Any surfaces that might abrade or damage the anchorage connector should be protected, padded or avoided (Fig. 6.19). Numerous options exist for protecting anchorage connectors.

Wire rope slings with a protective plasticized sleeve are an especially good choice in industrial environments. If a wire rope sling is not available, a sling made out of kernmantle rope may be a better choice than one of webbing because kernmantle rope of a given strength tends to be more abrasion resistant and usually stronger than webbing of comparable strength.

Straps designed for use as anchorage connectors often feature sewn terminations, or may even be sewn into a continuous loop. Webbing loops may be used either in a basket hitch (Fig. 6.20) or a girth hitch (Fig. 6.21) configuration to connect to an anchorage point.

Figure 6.18 Y-angle at tangent of anchor sling should not exceed 90°.

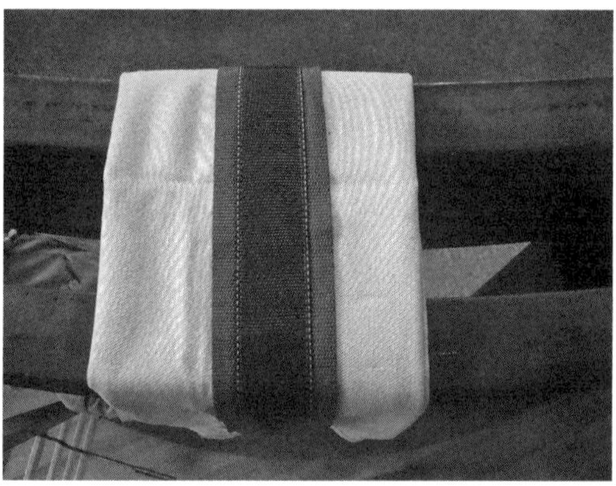

Figure 6.19 Use abrasion guards to protect anchorage connectors.

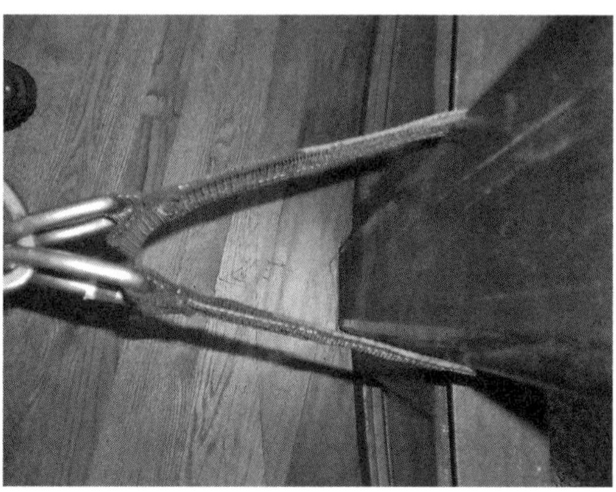

Figure 6.20 Basket hitch.

Figure 6.21 Girth hitch.

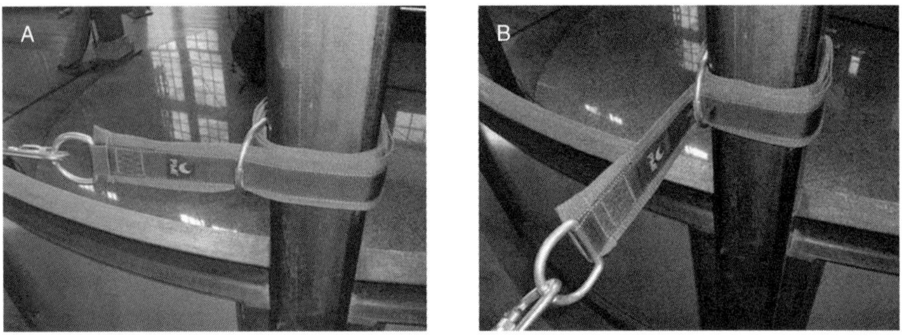

Figure 6.22 (A) Properly aligned girth hitch. (B) Poorly aligned girth hitch.

Some jurisdictions do not allow girth-hitched anchorages because an improperly aligned girth-hitched sling can be as much as 50% weaker than in a basket hitch configuration. This is not to say that a girth hitch should not be used, only that care should be taken to use it properly, so that the strength of the anchor is not reduced below the required system strength. Dictating rules for what *not to do* in rigging is generally not as effective as stating requirements for system performance, and then ensuring that personnel possess the knowledge and skills to create and analyze a system that meets those requirements. Proper use of a girth hitch involves aligning the load properly, as shown in Figure 6.22A, rather than creating a sharp bend in the sling where it exits the hitch, as illustrated in Figure 6.22B. It is also notable that some jurisdictions are reluctant to accept basket hitched anchorages because of the potential for cross loading connectors if the angle is excessive. Again, proper rigging can prevent this hazard.

OSHA requires that anchorages for fall protection be independent of the means of supporting or suspending the employee.[1] Industry best practice is that separate, independent anchorages should also be used for primary rescue systems, secondary ("backup") systems, and re-belays (re-anchors), all of which are discussed in Chapter 7.

[1]USDOL 29CFR 1910.66 App C (b); USDOL 29CFR 1926.502(d)(15).

In this case, the term "independent" means that each anchor is not reliant on another for security. This does not necessarily mean that the anchors are not connected to one another, or to a common system. For example, in some cases it may be advisable to pre-tension between a primary and a secondary anchor, or to use a primary anchor as a direction change for a backup line. This can aid in rope management, and does not detract from the fact that functions of the anchorages remain independent from one another. A *primary system* is considered that upon which the user is primarily reliant. A *backup system* is a system that is in place in case the primary system fails.

Satisfying OSHA's requirement for independent anchorages means that if the user is standing on a firm surface on which he or she relies for primary support, it is possible that only a secondary or "backup" system is required. If the user is relying upon roped systems for both primary support and backup safety, then two separate anchorages will be required, one for each system. Each anchorage may be comprised of a single highly reliable anchor point, or of multiple anchor points.

PERSONAL SAFETY

The first rule of rescue is to keep oneself safe. Personal safety is of utmost importance for any would-be rescuer; without this it is not possible to help others.

As discussed in previous chapters, anyone attempting to rescue another person should receive and adhere to safety training that is at least equivalent to that of the person(s) they are attempting to rescue. In the industrial environment, it is assumed that this means that the co-worker assisted rescuer will possess adequate fall protection training, equipment, and capabilities. This chapter will include superficial discussion on some key safety skills that every rescuer should have, but this is not to be considered a substitute for a site-specific fall protection plan and hands-on training by a competent trainer.

Clearly the best protection from falling is achieved when the rescuer simply avoids entering the fall-hazard zone altogether. The fall-hazard zone is comprised of all areas in which a person might be exposed to a potential fall. While the best way of avoiding a fall hazard is to stay out of the fall hazard zone, sometimes this is not realistic. Whenever a rescuer must approach or enter the vertical environment, some means of safety backup should be exercised. The principle of a safety backup system is to provide security in the event that the primary system fails.

A safety backup system for co-worker assisted rescue should provide protection that is at least as effective as OSHA fall protection requirements, should allow the primary system to function without interference, and should provide safety against catastrophic results in the event of a primary system failure.

Positioning or restraint techniques are often a good choice for use by co-worker assisted rescuers. Restraint systems are those that effectively serve as a "leash" to prevent the user from entering an area in which he might be exposed to a fall. When using fall restraint, the Authorized Person is secured to an anchorage using a lanyard short enough to prevent the person's center of gravity from reaching the fall hazard, but long enough to allow the person to perform necessary work. These systems are described in greater detail in Chapter 1.

In preplanned systems, a mechanism for fall protection may already exist. For example, ladders are often fitted with safety systems, such as a safety cable or rail. Areas where work at height is frequent or common may have self-retracting lifelines in place. Whether working from a ladder or ascending the structure itself, rescuers should be protected against a potential fall.

Access and Egress

The safety of the rescuer cannot be overemphasized. If possible, accessing the subject using a stairway, platform, or other stable and secure means is always the best choice. However, the fact of the matter is that rescuers will seldom find the subject in the most convenient of locations. Often the subject will be suspended in the most awkward predicament, unreachable from common walking and working surfaces.

The methods in this book preferentially recommend use of "non-entry" methods for rescue—that is, methods that permit extrication of the subject without the rescuer having to enter the vertical or hazardous environment. Sometimes, however, a rescuer must ascend or descend to gain access to the subject.

Free Climbing

Although the term free climbing[2] has sometimes erroneously been used to describe the idea of unprotected ascent of a structure, in its purest sense the term *free climbing* simply refers to the process of using one's arms and legs to "climb" a structure, unrestricted and unconfined by a specific path and using the structure itself for support.

A free climber who elects not to use a rope or any protection at all is said to be "free soloing." While certain industries and applications may have allowances or exceptions that permit unprotected (also called free) climbing, this is never a recommended approach, especially in times of duress such as during a rescue. In both work and rescue activities, 100% fall protection should always be the goal. Use of protective devices, including vertical lifelines, twin lanyards, or belaying, is highly recommended.

Ladders

In some situations, the rescuer may be able to ascend to the subject utilizing a fixed ladder. Climbing a fixed ladder should be treated much like climbing a structure; that is, the climber should be protected from a fall.

[2]This reference to free climbing should not be confused with the way in which the term is used in recreational climbing. In recreational climbing, the term "free climbing" is used to depict a climber ascending a face, even while roped. In industry, especially as relates to the Gannett Variance, the term "free climbing" is used to express the idea of a climber ascending a structure, such as a ladder or lattice, without rope or lanyards.

Figure 6.23 Ladder with cage.

Some ladders are equipped with a *ladder cage*. Ladder cages (Fig. 6.23) are common at industrial worksites because there is an OSHA regulation requiring them on fixed ladders that are over 24 feet in height. However, it can be argued that ladder cages pose additional potential hazards to the user, and do little more than create a false sense of security.

A ladder cage typically starts several feet above the lowest rung, and essentially surrounds the ladder so that the climber must pass through the cage as he ascends the ladder. The purpose of this cage is ostensibly to protect the worker. However, the ladder cage does not effectively prevent or arrest a fall, it merely restricts the movement of a worker falling backwards. Rescuers who must climb a ladder should use fall protection other than a cage, as described in Chapter 1, to protect against a catastrophic fall.

In the best of circumstances, the rescuer may find the ladder to be equipped with a vertical lifeline (safety rail or cable) that is purpose designed to protect the climber from a fall. However, if the ladder is also equipped with a cage, the cage can prevent the vertical lifeline from functioning correctly. Extra precautions should be taken in this case to protect against a fall.

Rail systems are typically mounted in the center of the ladder. Using a special grabbing mechanism that is specifically designed to lock onto the rail, the climber ascends the ladder while connected to the rail. The grabbing mechanism self-trails as the climber ascends, but locks onto the rail should the climber fall. Cable systems function in much the same manner. The cable may run alongside or the center of the ladder, and is designed to be used with an appropriate cable-grab device; there are numerous models of these, and it is important to verify compatibility between the cable and the cable grab.

The rail grab and cable grab devices used in these systems are very system-specific. For best performance, the correct, manufacturer-specified device must be used with each system. If rescuers anticipate needing to access a cable- or

Figure 6.24 Damage to ladder or safety cable may not be visible until after you begin climbing. Credit: JP Jones, Safety LMS.

rail-protected ladder, arrangements should be made in advance to ensure a sufficient supply of the appropriate grab devices that fit that system. During a rescue it seems that there are never enough cable grabs or rail grabs available, and the temptation to use a grab that is designed for a different system should be avoided!

A potential hazard of both the cable and rail systems is that activation of the locking mechanism is highly dependent upon the direction of pull. If the climber should fall but for any reason remain supported in an upright configuration, and/or if the device is not subjected to the full force of the body, there is some risk that the locking mechanism will not engage. This is the hazard that can be created by a ladder cage, and can result in a subject being wedged awkwardly inside the ladder cage. This makes for a very difficult rescue.

Another hazard involved with ladders and pre-installed safety cables is the inability of the rescuer to inspect the system(s) before he or she climbs. Ladders and cables are usually anchored at the top, and perhaps at intermediate points, with bolted or swaged terminations (see Fig. 6.24). The only way to ensure that this termination is secure is to inspect it directly—which requires the user to first climb the ladder. Until one has verified that the cable, including termination, is secure, it is best to use an alternative method of safety.

Lack of a fixed safety system does not mean that it cannot be climbed safely. One way to climb an unprotected structure or fixed ladder is to use twin lanyards with large "ladder hook" style connectors and force absorbing device(s), which were first introduced in Chapter 1, the section on "Fall Arrest Using Horizontal Lifelines." As the climber ascends, he or she clips one leg of the lanyard as high on the structure as he or she can reach, then climbs up a step or two, and attaches the other leg higher on the structure before removing the previous lanyard leg (see Fig. 6.25).

When climbing a ladder using this method, you may clip directly to the rung as long as you are sure that it is sufficiently strong. Alternatively, try clipping along the outer rail of the ladder, taking care to position the hook so that the gate will not

Figure 6.25 Climbing using twin lanyards.

be cross-loaded in the event of a fall. Climbing a ladder using twin lanyards is a very physical task, and can be exhausting for a rescuer—especially if he or she is carrying additional equipment—but this is one means by which you can ensure 100% fall protection, without compromise.

It is worth noting here that some fixed ladders are equipped with climb-assist devices. A climb-assist is a mechanized device that is intended solely to relieve the climber of some of the exertion that would otherwise be required for climbing by exerting an upward force on the climber's harness. The climber simply connects the climb assist device to his sternal attachment point and begins to climb. As he or she ascends, the climb assist exerts an upward force, relieving some of the effects of gravity on the climber. A climb-assist is generally *not* also intended for use as protection against a fall, so take care to use additional fall protection as necessary.

Belaying

Belaying is one versatile type of safety system that may be used when climbing almost any structure, including a ladder. Belaying involves one person (called the belayer) having some control over a device and rope to which another person (the rescuer, in this case) is connected. As the rescuer moves, the belayer pays out or takes in rope to minimize the potential fall distance. The belayer must take care not to tug on the rescuer or pull him or her off his or her stance.

To use a belay system:

1. Connect a belay device to an appropriate anchor.
2. Thread the rope through an appropriate belay device.
3. Attach the working end of the rope to the rescuer's harness.
4. Belayer feeds and/or pulls the rope through the device as the worker moves, taking care to avoid both excessive slack and excessive tension.

A climber may be belayed from above or from below. Belaying a climber from above generally provides greater security, but of course it is only possible if the belayer is positioned at a level higher than the climber. The first person to ascend a structure can be protected, but he or she may be exposed to additional fall protection as he or she must place his or her own protection as he or she progresses and then climb above the last anchor point in order to proceed. This is known as *lead climbing*.

Lead climbing is a learned skill, and is a good way to access otherwise unprotected heights. While the concept stems from practices found in recreational climbing, this technique has professional applications, particularly where protection to a worker may otherwise be deemed "infeasible." The practice may be used to protect a ladder, to climb a tower, or to climb almost any structure.

The lead climber must trail his or her protection rope, which is managed by the belayer at ground level. The climber climbs a few feet, places a sling, bolt, or other protection, clips into it, and then climbs a few more feet to place a new anchor; and so he or she proceeds.

Choosing this method of protection adds an element of mobility while still allowing the rescuer (or worker) to maintain 100% fall protection. It should be noted that a lead climber will be exposed to a fall potential, sometimes perhaps even in excess of fall distances specified in regulatory language. A decision to take this calculated risk should be undertaken only with careful attention given to avoiding harmful collision with obstructions, and to reducing potential impact forces for the protection of the worker.

Ascending

It is also possible to ascend to a high point on a structure using a *fixed rope*. The term "fixed" here does not denote a rope that has been repaired, but rather one that is stationary, or in a fixed location. Of course, this does require first establishing a fixed rope in place, which may be accomplished by the first person lead climbing the structure itself. Once the fixed rope in place, ascending is a quick and secure way of getting multiple people up a structure relatively quickly.

Ascending a fixed rope, illustrated in Figure 6.26, is a relatively simple matter if the rescuer is equipped with suitable equipment. The foundation of a climbing system lies in rope-grab mechanisms, such as personal ascenders, that are designed for this purpose. Note that ascenders are a special type of rope grabbing mechanism, and are not the same as rope grabs intended for use in fall arrest. The two are not interchangeable.

Rope grabs/ascenders that are designed for climbing rope (rather than for fall arrest) typically have a more aggressive cam, are less likely to slip while in use, and are more easily attached to and adjusted on the rope. These are all desirable characteristics of ascenders, but undesirable characteristics of a fall arrest rope grab.

When ascending a fixed rope, the climber should always maintain at least two points of contact to an appropriate suspension attachment, such as a sternal or waist D-ring. In addition, at least one foot loop will be required to assist with upward progression. An example of an appropriate ascending setup is shown in Figure 6.27.

Figure 6.26 Ascending a fixed rope.

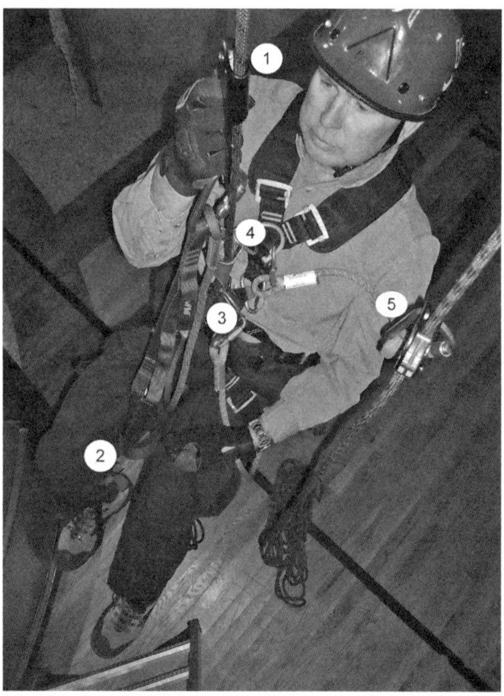

Figure 6.27 An ascending system for ascending a fixed rope should include (1) handled ascender, (2) footloop from handled ascender to foot, (3) safety connection from handled ascender to harness, (4) progress capture (chest) ascender, and (5) backup safety system.

To ascend using this type of system:

1. Don your ascending system as shown above.
2. Attach your chest ascender to the fixed rope according to the manufacturer's instructions.
3. Connect a footloop or etrier and a waist lanyard to your handled ascender and attach the handled ascender to the fixed rope according to the manufacturer's instructions. (Note that the waist lanyard should be attached at a length that allows you to extend your arm almost completely, leaving the footloop in a comfortable stepping-up position.)
4. Slide your handled ascender as high as you can on the rope.
5. Step into the footloop that is attached to the handled ascender and stand up, sliding your chest ascender as far as possible up the rope.
6. Sit down until your weight is primarily on your chest ascender.
7. Lift your foot to release the tension in your footloop and slide the handled ascender up.
8. Repeat the process until you have ascended the rope as far as you wish.

There are many other types of appropriate systems for ascending, each suited to different types of equipment, environments and needs. If you intend to ascend rope often for the purposes of work, certification in rope access methods and/or specialized training in ascending methods is highly advisable.

Descending

If the rescuer is above the subject when help is required, it may be necessary to descend instead of ascending to get to the subject. Again, descending via stairs or a fixed ladder would be the preferred choice, but sometimes this is not feasible. Rope-based techniques for accessing a subject from above may involve being lowered from above (addressed in later chapters), or by descending. All things being equal, and if sufficiently trained personnel are available, being lowered is often a better choice than descending because the system lends itself to being converted to a raise if necessary; lowering and raising systems are discussed in further detail in Chapter 9.

Descending via rope is addressed in this chapter as a personal skill that all workers at height, and certainly all potential co-worker assisted rescuers, should be capable of performing. Shown in Figure 6.28, this technique provides a foundation for access, work positioning, personal escape, and partner rescue. Descent techniques used for personal escape and co-worker assisted rescue should reflect equipment and methods that are intended for workplace use. In other words, with few exceptions, recreational climbing/rappelling methods are generally not suited for workplace applications. Not only should equipment be purpose-designed, but a Qualified Person should ensure that all components that will be used together are compatible and of sufficient strength and performance for the intended use.

Descending techniques require the user to be highly trained and competent, and to be aware of the limitations of such systems. An understanding of the principle

Figure 6.28 Descending a rope. Credit: Pigeon Mountain Industries, Inc.

of double protection, that is a working line plus a safety line, is essential for users. Given OSHA's requirement that the backup system be independent of the primary system of support, rescuers should clearly understand that descending systems and techniques should always be used in conjunction with a safety line and backup device unless the use of such secondary system is likely to create greater hazard than it protects against. It is also understood that during personal escape in an actual emergency, a secondary or backup system will not likely be available. This should not preclude the use of a single line for emergency egress!

Descenders exist in a variety of forms, and may function quite differently one from another. At the most basic level, all descenders offer a similar function: that is, they provide friction when used with a rope, for the purpose of slowing or stopping a load from traveling downward with gravity. The International Standards Organization (ISO) classifies descending devices by type. There are six different types identified, as described in Chapter 4. It is possible for a single descending device to conform to the requirements of more than one type.

Some descenders, called controlled rate descenders or CRDs, do not require human operation; that is, they function automatically, lowering the suspended user at a fixed rate. This type of descender usually consists of a centrifugal brake and a wire or rope line, and most often the descender portion of the system is fixed in place.

Figure 6.29 Using a controlled rate descender.

To use a controlled rate descender, simply clip the end of the line to your harness and step out into the ethers (Fig. 6.29). As soon as the descent line is sufficiently weighted, descent begins. Although these are relatively easy to use, it is this same simplicity that creates limitations when it comes to being able to raise, stop, or alter the direction or motion of the load.

If the descender is one that requires human operation, it is critical that the person operating the descender always keeps the device within reach. In systems where the device travels with the user, the descender should normally be attached directly (usually via a connector) to a waist or sternal attachment point on the user's harness.

For every descent, a decision must be made whether to drop the free end of the rope down the vertical plane prior to descent, or to carry it. In either case, put a stopper knot or some other blockage onto the end of the rope to prevent accidentally descending off the end.

If the distance is long or the rope is heavy, dropping the free end ahead of one's descent may be the easiest solution. Exercise caution when doing so to avoid tangling or catching the rope on any obstructions. Note that the weight of a great deal of rope below will create tension on the braking side of the descender. You must manage this during descent by lifting the rope so that it does not impart so great an effect on the descender.

There is a method to dropping or throwing the free end of a rope over an edge, and you should practice this method before attempting to do it in an emergency.

First, select where you would like for the rope to hang. It can be quite difficult to move after it has been thrown. Be sure to select a rope path that is near the subject, but not so close that the subject can reach the rope. If the subject can reach the rope, he or she may be tempted to try to grab it, which can create some serious problems. First, the subject may endanger him- or herself in flailing around trying to grab the rope. Second, if the subject is able to take hold of and pull on your rope while you are descending the extra friction can stop your descent. Finally, take care to not hit the subject when you throw the rope.

Before throwing the free end, be sure to anchor the standing end of the rope to an appropriate anchorage; do so without encumbering or entangling the working end of the rope. Then, stack excess rope on the ground next to the edge—close enough so that it will feed freely but is clear of any obstacles (including your feet)—with the standing end at the bottom of the stack and the working end of the rope coming off the top of the stack. Pick up the working end of the rope, and coil it gently in your hands. Draw six to eight loops of rope into each hand, with the first six to eight loops nearest the working end in your throwing hand, and six to eight additional loops in your other hand.

The person throwing the rope should be secured by some means to prevent him or her from accidentally launching him- or herself along with the rope. Before actually throwing the rope, yell "Rope!" loudly enough so that those below can hear. Pause briefly (to give people time to get clear), holler "Rope!" again, and throw—first the loops from your throwing hand, and then tossing the other loops after, and finally letting any remaining rope feed from the stack on the ground.

An alternative solution to throwing the rope is to carry the free end of the rope with you in a bag, feeding it out as you descend (Fig. 6.30). While the bagged

Figure 6.30 Descending with a bagged rope.

Figure 6.31 Transitioning over a 90° edge. Credit: Vertical Rescue Solutions/PMI.

rope may be heavy, the bag can be connected to a utility ring on your harness or even to the same connector as your descender and/or seatboard. This method helps to avoid potential rope entanglement, avoids any hazards associated with the subject being able to reach the rope, and is especially advisable during emergency escape/egress.

Edge Transitions Getting over the edge is the most difficult part of any descent. In general, a sloping edge will be easier to negotiate than a sharp one, and an undercut edge will be especially difficult to get over. This is because your feet are integral to keeping your body away from the surface you are descending, and with an undercut edge there is no place to put your feet. To help facilitate a smooth edge negotiation, position your anchor as high as possible in relationship to the edge (at least waist height, and preferably higher) and ensure that it is placed back from the edge.

To descend over a 90° edge, simply tension your descender at the edge of the drop, and climb down over the edge until you have fully loaded the rope, as illustrated in Figure 6.31. Of course, this is not the most graceful method, but it does help prevent falling and impact loading your systems.

For a more graceful entry, stay in an upright position, feet just at the edge of the drop, while feeding rope through the descender and leaning backward into the system. With your legs extended shoulder width apart, allow your derriere to precede your feet and bend at the waist until you are in a seated position, with your feet at

Figure 6.32 Standing transition over an edge.

about waist level (Fig. 6.32). This is roughly the position that you will maintain throughout descent.

When transitioning over a very difficult edge, such as an undercut one, it may be necessary to descend until your feet are above your harness attachment point, perhaps even as high as your head, before moving your feet. As a general rule, keeping your stance wide and your legs high offers the greatest stability. Do not jump, bounce, bound, or use other uncontrolled methods.

The above methods take practice and skill. Alternatives for negotiating a difficult edge include sitting on the edge and scooting yourself over gently (Fig. 6.33), or fastening an etrier or footloop at the edge and using it to downclimb until your system is tensioned (Fig. 6.34).

Communications during Descent

In practice, a safety officer (who may also be called "safety attendant") or second rescuer should usually be stationed at the top of any slope or structure where someone is descending. While a rescuer is always primarily responsible for his or her own safety, the role of the safety officer/attendant is to double check all aspects of the system, and to monitor the system as it is loaded to help ensure safety. The safety attendant may also serve as a belayer if this is the method being used by the rescuer for backup. Communication between the safety officer and/or belayer should be crisp, effective, and to the point. If voice communications are used, radios may be necessary to compensate for long distances or noisy environments.

When a rescuer is completely connected to the system, he or she should allow the safety attendant to double check the connections and the system as a whole. If

Figure 6.33 Sit on the edge and scoot yourself over.

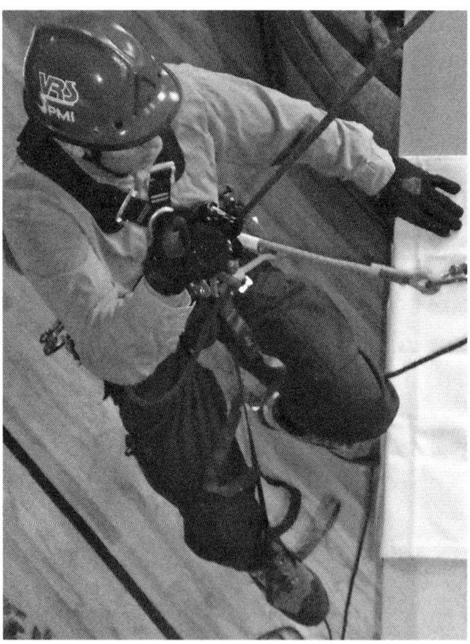

Figure 6.34 Use an etrier to downclimb until your system is tensioned.

he or she is preparing to be belayed, the following series of commands can be used to communicate:

RESCUER: On belay?

BELAYER: Belay on.

RESCUER: Descending.

BELAYER: Okay.

RESCUER (upon reaching the bottom and clearing the system): Off belay.

BELAYER: Belay off.

To descend using a device that travels with the user, follow these steps:

1. Don an appropriate harness with a waist or sternal attachment.
2. Connect your descender to the waist or sternal attachment with an appropriate connector.
3. Connect the standing end of the rope to an appropriate anchorage.
4. Thread the life safety rope through your descending device (see your specific device instructions for further information).
5. Verify that the descender is properly connected to your harness or evacuation seat using a locking carabiner.
6. Attach a belay, vertical lifeline, or other secondary backup system to your harness. Note: It is acceptable to use the same harness attachment as for your primary system.
7. Position yourself at least two feet from the edge over which you intend to descend.
8. Facing the anchor, pull the rope backward through your descender to remove any slack, creating tension between your harness and the anchor.
9. If you intend to carry the rope with you as you descend, while still facing the anchor attach the bag to your system as appropriate; if you intend to drop the free end of the rope down ahead of you, do so.
10. Check that the area of descent is clear (and, if applicable, that the entire length of rope is free from obstruction, tangling, or hazards). Ask the safety officer or belayer to double check your system.
11. Once the safety officer or belayer has confirmed readiness of the system, ask “On belay?” to your belayer, or announce to the safety officer “Ready to descend.”
12. On the affirmative reply from your belayer or safety officer (“Belay on” or “Descend,” respectively) gently lean into the system so that your weight is supported by the rope, connected to the anchor. At this point you should be leaning into the system but still on a safe surface, not yet over the edge.

13. Hold tension on the tail of the rope with your brake hand and, if using an auto-locking descender, activate the operating mechanism.

14. Relax tension on your brake hand until you are able to allow rope to travel through the descender.

15. Walk backward toward, and then over, the edge, adjusting your hand tension to maintain full control of your rate of descent at all times. *Never let go of the rope with your brake hand!*

16. Negotiating the edge is the most difficult part of descending. See the Edge Transitions section for discussion. As you negotiate the edge, take care to control the path of your rope as close to the fall line as possible. Lay the rope where you want it as you go, and pad it as necessary (see Fig. 6.35). Once the loaded rope is pressed against the surface, it is very difficult to move and is subject to abrasion or damage.

17. Once over the edge, "sit" in your harness, legs extended forward near waist level. Keep your legs straight, but without locking your knees. A stance with feet shoulder width apart will offer good control.

18. Descend steadily by "walking" backward down the slope or structure. Maintain balance and control, taking care to avoid abrasion, sharp edges, falls, bouncing, or other hazards.

19. Once you have reached the bottom, unclip your harness connection from the descender, notify your belayer/safety officer that you are clear ("Off belay" or "Off descent," respectively), and move to a safe location.

Figure 6.35 Protect sharp or abrasive edges as needed. Credit: Pigeon Mountain Industries, Inc.

PERSONAL ESCAPE

As has been alluded to in other chapters, the foundational rule of responding to an emergency is to not become or create an emergency oneself. Foundational to this is the need to maintain good work practices and avoid accidents, but it also means being capable of extricating oneself from a stuck position or from an arrested fall. Even a person working atop a platform or in a bucket truck should have some means of personal escape, in case the usual means of access/egress is cut off or rendered inoperable for some reason. That need may take on a greater sense of urgency in the event of a fire or other sudden arrival of a hazardous condition.

While a ladder, mobile stairs, or lift truck are all acceptable means of personal escape, at-height workers should also be adept in the use of an emergency escape descending system. There are a variety of these available on the market, and at a minimum they usually include a rope and a descender. It is usually recommended to use a manufacturer-designed kit rather than piecing one together from components made by different manufacturers, so as to be sure that the components are compatible. Emergency escape, then, simply becomes a matter of being able to put these devices into action on demand.

Descending techniques described earlier in this chapter provide the foundation for emergency escape by descent. These skills should be practiced periodically, at a minimum once or twice per year, by anyone who might use them. Practice should be performed using equipment identical to that which the individual might use, although each person's actual emergency kit is best kept in new, unused condition to help ensure that it will be intact and operable when needed.

Practice should take place in the actual environment, or at least the same *type* of environment(s), where an individual might be expected to perform in an emergency. This will help prepare the individual for selecting anchors in that environment, and for managing the specific hazards and conditions associated with it.

Although an actual emergency escape will generally occur on a single-line system, without a belay, it is advisable to belay workers while they are practicing these techniques, especially if they are new to the equipment, new to the environment, or new to the employer. Skills practice should always be overseen by a skilled technician (preferably a competent instructor) who will ensure that everything is monitored and double checked during the practice session for safety purposes. Provision for rescue should be made in advance in case of a mishap during practice.

Practicing Emergency Escape from a Platform

1. Select an appropriate anchor with a strength of at least 3,000 lbf (or greater if required by jurisdictional authorities).
2. Connect the standing end of the rope to an appropriate anchorage, taking care to ensure that the anchor:
   - Is protected from abrasion
   - Is rigged so as to avoid rollout
   - Is as high as practicable—preferably at least waist height

- Will load properly when weighted
- Is in line with the fall-line

3. Ensure that the descending device is properly rigged to the rope, and that it is close enough to the anchor to prevent uncontrolled loading when the system is loaded.
4. Connect the descending device to your harness with an appropriate connector.[3]
5. Connect to the belay.
6. Lean into the system in the same direction that you will be descending, but without putting yourself in a fall hazard.
7. Double check all personal protective equipment (PPE), and verify that the system is properly rigged.
8. Attach the rope bag to your harness on the same side as your brake hand, or drop the end of the rope down the fall line, in accordance with your organization's prescribed practices. Ensure that there is a stopper knot at the end of the rope, to prevent descending off it, and that the rope is not otherwise knotted or tangled.
9. Feed rope through the descender, making sure before you go over the edge that it is functioning properly.
10. As you transition over the edge, avoid impact (shock loading) on the system.
11. Pad abrasive or sharp edges, and descend.

While it is important to practice in environments that replicate real work situations (Fig. 6.36), avoid exposure to potential workplace hazards such as machinery, energized lines, high winds, and obstructions. During an actual emergency, pay particular attention to such conditions and do your best to avoid putting yourself in danger. In an emergency, there may be multiple options for where to descend. Consider alternative routes for maximum safety, and be sure before you descend that your rope will reach the ground—even if you plan to carry the bagged rope with you. If possible, activate additional emergency response, whether an in-house rescue team or external, municipal rescue, before attempting personal escape.

Practicing Emergency Escape from Fall Arrest

Recovery from fall arrest into a personal escape system is a more complicated proposition, especially for workers utilizing a dorsal attachment for fall arrest. It is quite possible to fall into an arrest system and be completely uninjured, and in this case you should consider self-extrication; in fact, not extricating oneself can lead to greater injury. Studies have shown that it can be quite hazardous to remain suspended

[3]Use of a harness with a front attachment point (either waist or sternal) is required for emergency escape. If you are not likely to be wearing such a harness during work, consider keeping a harness or emergency seat as described in Chapter 4 with your emergency escape kit.

Figure 6.36 Practice emergency escape in actual work environment.

for prolonged periods in fall arrest (Fig. 6.37), as suspension intolerance can create problems in otherwise healthy individuals.[4]

Attaining a Position of Comfort The moment a person falls into a fall arrest system, local emergency protocols should be initiated, including co-worker response plans as well as activation of municipal responders, if applicable. A person who has fallen into a fall arrest system should always be checked out by medical professionals as soon as practicable.

A person suspended after an arrested fall will typically be left hanging from either the dorsal or sternal harness attachment point. The first step in recovering from this situation is to attain a position of comfort as soon as possible. Stepping into a footloop will allow one to relieve compression points in one's harness and take some pressure off of the diaphragm, so that one can breathe more easily (Fig. 6.38). This is more readily accomplished, and with better results, when the subject is suspended from the sternal attachment rather than when he or she is suspended from the dorsal attachment.

A footloop can be as simple as a loop of webbing, or may be a series of "steps" at different levels to facilitate greater mobility. There are many specialized products

[4]Nina L. Turner, James T. Wassell, Richard Whisler, and Joyce Zwiener, "Suspension Tolerance in a Full-Body Safety Harness, and a Prototype Harness Accessory," *Journal of Occupational and Environmental Hygiene*, 5(4):227–231 (2008).

Figure 6.37 Worker suspended in fall arrest.

Figure 6.38 Using a footloop to attain a position of comfort.

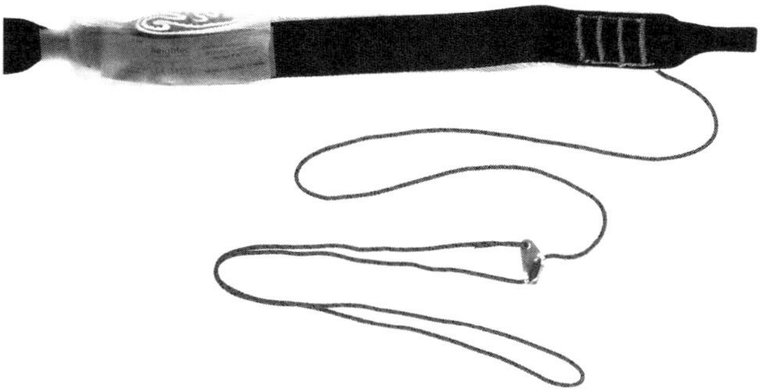

Figure 6.39 Self-deploying suspension relief strap. Credit: Heightec.

commercially available for this purpose. Some harnesses even include suspension relief footloops that can be deployed after a fall.

Whether you are using a simple footloop or something with multiple steps, consider how and where the suspension relief strap will be attached to your system. Some suspension relief devices are designed to attach to the upper/chest portion of the harness, but this can create a relatively ineffective condition where the user is pulling against his or her own body rather than the anchored system.

It is much easier to use a footloop that is connected directly to the suspension system in a manner that does not exert oppositional force on your body. For example, one commercially available force absorber is designed to automatically release a footloop when activated, with the result being that the footloop extends directly from the lifeline rather than the harness (Fig. 6.39).

Personal Escape from Suspension Having attained a position of comfort, you now have some time to calm down and plan your escape from your suspended condition. This will ideally have been preplanned, and may involve deploying a personal escape system, or it may mean simply waiting for co-worker assistance.

If deployment of a personal escape system is the preferred plan, be sure you have practiced this often. The ideal emergency escape system will be lightweight, pre-rigged, and contain an anchorage connector, rope, and descender with connector, all in a small bag (Fig. 6.40).

To deploy a personal escape system from a fall arrest condition, follow these steps:

1. Before attempting personal escape, notify someone of the situation and request that additional resources be activated (they can always be stood down later, if not needed).
2. Find a secure, reachable anchorage point for your escape system.
   - If you are working against a structure, it may be possible to use a part of that structure as your anchorage as shown in Figure 6.41.
   - In a free-hanging situation, it may be necessary to anchor your rope directly to the line you are already hanging from. If you have fallen on the system

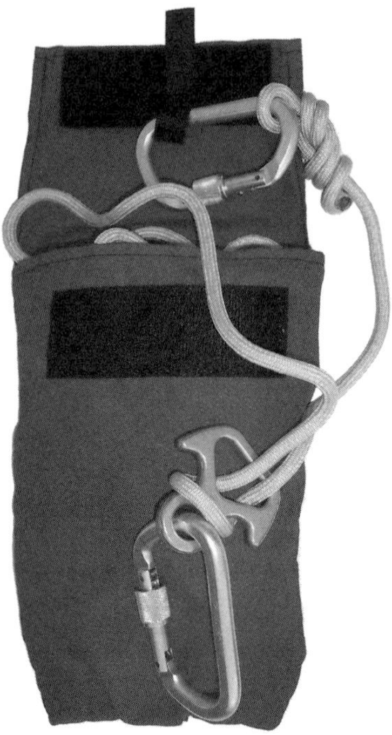

Figure 6.40 Personal escape system. Credit: Pigeon Mountain Industries, Inc.

Figure 6.41 Connect to the structure if possible.

Figure 6.42 If you must, connect directly to the fall arrest system from which you are hanging.

already, be sure that neither the rope nor the auxiliary equipment that you will be using is compromised.

3. Connect your anchorage connector to the anchorage point you have selected.
   - In a free hanging situation, additional agility may be required to connect your descending system to the rope grab or other component of the fall arrest system you are hanging from (Fig. 6.42).
4. Connect the descender of your emergency escape system to your sternal or waist attachment point.
5. Stand up in your footloop or other method to release the tension from your fall arrest system, as shown in Figure 6.43. Simultaneously tension your escape descender. This may take several small steps to fully accomplish.
6. Once your full weight is on the escape system and the fall arrest system is no longer under tension, disconnect yourself from the fall arrest system so that you may descend. *Take care at this time to* not *compromise your escape system, which is now your* primary *support!*
7. Make sure that your rope is long enough to allow you to reach safety.
8. Descend slowly, maintaining control at all times (Fig. 6.44).
9. Seek medical attention immediately.

When attempting emergency escape from fall arrest, always begin by first attaining a position of comfort, and only use techniques that you know and have

Figure 6.43 Step into a footloop to help tension your descender.

Figure 6.44 Personal escape from fall arrest.

practiced. Avoid wearing yourself out completely—save a little energy in case you have to wait for rescue. Never attempt to cut yourself out of a system with a sharp instrument, and do not unclip anything without first ensuring that you are connected to an alternative means of safety.

POST-CHAPTER QUIZ

Chapter 6

1. The reason it is important to practice emergency rescue is
   a) To help ensure a better outcome in the event of a true crisis
   b) If you practice and are ready, it will never happen
   c) To avoid regulatory citations
   d) So everyone will think you are an expert
2. True or False: Rescuing someone you know well is usually much easier than rescuing a complete stranger.
3. Responders should only attempt to perform a skill that is beyond their level of skill and experience
   a) When it is a life and death situation
   b) When they have seen it done on TV
   c) When someone is available to talk them through it
   d) Never
4. The foundational rule of responding to an emergency is to
   a) Not become or create an emergency oneself
   b) Carry as much equipment as possible
   c) Dial 911
   d) Only use ANSI Z359-approved equipment
5. Some advantages of factory-sewn and -swaged terminations include
   a) Consistency and reproducibility
   b) Repeated point stress in a single location
   c) They limit the variable functionality of the system
   d) All of the above
6. The main purpose of a safety/backup system is to
   a) Provide primary safety to an Authorized Person
   b) Become entangled with the primary system
   c) Provide safety against catastrophic result in the event of a primary system failure
   d) Ensure regulatory compliance

7. A person who has fallen into a fall arrest system should *first* attempt to:
   a) Attain a position of comfort
   b) Wait for help from a co-worker(s)
   c) Use a personal escape system to descend to safety
   d) Use a personal escape system to ascend to safety
8. When practicing emergency escape, take care to ensure that the anchor
   a) Is protected from abrasion
   b) Is rigged so as to avoid dynamic rollout
   c) Will load properly when weighted
   d) All of the above
9. One way to make it easier to negotiate an edge during descent is to
   a) Keep your body as close as possible to the surface you are descending
   b) Select an anchor that is as low as possible in relation to the edge
   c) Select an anchor that is relatively high in relation to the edge
   d) Use a jumping action
10. To help prevent accidentally descending off the end of a line, a person can
    a) Place a stopper knot in the end of the rope
    b) Use a bagged rope, with the end of the rope securely fastened to the bag
    c) Either a or b
    d) Neither a nor b

ANSWER KEY

1. a
2. False
3. d
4. a
5. a
6. c
7. a
8. d
9. c
10. c

PART III

PUTTING IT ALL TOGETHER

CHAPTER 7

PRINCIPLES OF RIGGING

Upon completion of this chapter, you should have a firm understanding of:

- The positive and negative effects of gravity toward rigging.
- How friction and angles affect the rescue system.
- How to perform lowering/braking operations to facilitate co-worker assisted rescue.
- When and how to use mechanical advantage systems for raising a load.
- How to implement a belay system.
- The importance of and techniques for edge protection.
- Load sharing anchors.
- How to use directional deviations.

Using ropes for co-worker assisted rescue is a challenging and diverse subject, encompassing concepts that stem from basic physics and engineering concepts, climbing and caving techniques, and mechanical principles. Michael G. Brown, in his 1996 book *Engineering Practical Rope Rescue Systems*, likened the topic to "both the Hubble Space Telescope and a powerful electron microscope—the more you look both inward and outward, the more you see."

In keeping with the organizational format of this book, this chapter expands on the basic information and practical knowledge introduced in Chapters 4–6. It lays the groundwork for understanding rigging concepts that will be further expanded into rescue skills in Chapters 8 and 9. Here we will explore the concepts behind rescue systems and methods, while Chapters 8 and 9 provide more detailed guidance in specific techniques.

At the most foundational level it can be said that much of the important understanding of rope rescue equipment, skills, and techniques is grounded in two key principles: *gravity* and *friction*.

GRAVITY

From the moment of our arrival on earth we exist in a world where gravity rules. Gravity is more or less constant on our planet, one of the few natural phenomena that we can somewhat accurately predict. Gravity refers to the acceleration that a

Falls from Height: A Guide to Rescue Planning, First Edition. Loui McCurley.

body, in this case the Earth, imparts to objects on or near its surface. Gravity on earth is calculated as 9.81 m/s^2, although science has learned that actual gravitational forces vary over different parts of the globe. Sometimes we call gravity a "law," but this insinuates more control than we actually possess. Really there is nothing we can do to force the universe to function according to the principles of gravity—we're just fortunate that it does!

The force of gravity being relatively consistent here on earth, eventually we learn to overcome it and a short while after birth we begin to crawl. When we first stand up, we fall down. As our muscles strengthen we learn to combat the effects of gravity, but our feet remain stuck to the ground. Through creativity we have invented methods to reach greater heights—stairs, ladders, elevators, hot air balloons, airplanes, and even ropes and cables.

All the while gravity still exists in our world as a relatively constant, predictable force. Knowing this, we can learn to use gravity to our advantage in work and rescue at height, just as we do in other aspects of life.

A ball rolls down hill. That's gravity. Contact with the hill imparts friction. This affects direction and speed of travel of the ball. Terrain features also have an effect. A ball rolling down a hill where friction and terrain features affect its path follows the path of least resistance. This path is called the fall line.

Any load suspended at height will similarly follow the natural gravitational pull, or fall line, affected and redirected only by opposing forces in its path. Certainly other factors are at play but gravity is constant and predictable, and friction/terrain are largely influential.

With these concepts in mind we can choose from where, and how, to suspend our rescue load in order to best utilize gravity to our advantage, and we can use the effects of friction to help control its path and rate of movement. It is easier to learn to use gravity to our advantage than to fight it.

FRICTION

Work is defined as the exertion/expenditure of energy. In rescue, our goal is to direct that energy in predictable and productive ways in order to achieve a goal. Nearly all of the "work" performed during rope rescue is managed and directed using some combination of friction and angles.

Friction is defined simply as "The resistance of an object to the medium through which or on which it is traveling."[1]

In rope rescue, friction on the rope occurs in relation to other equipment as well as the environment. When planned into the system, friction can help the rescuer control a load, and can contribute positively to the operation. When not planned for, friction can hinder the rescue, make the rescuer's work harder, or even damage equipment and jeopardize the operation. Generally speaking, friction is our friend when lowering a load, and our enemy when raising.

[1]"Friction," *The American Heritage® New Dictionary of Cultural Literacy,* 3rd ed., Houghton Mifflin Company, 2005; http://dictionary.reference.com/browse/friction (accessed June 21, 2011).

Perhaps the most common and least controlled friction mechanism in rescue is the edge(s) over which the rope bends from a near horizontal to a near vertical orientation when a load is suspended (Fig. 7.1).

The specific nature of this edge will determine the effect that the edge has on the rope and on the system. In terms of system efficiency, a smooth edge is generally better than a sharp edge, and a gentle bend is preferable to an extreme one—although the distance over which the rope comes into contact with the edge will also play a role. Friction not only affects how efficiently the system performs, but can also have a damaging effect on equipment.

Modern kernmantle life safety ropes offer good protection from frictional damage, but additional protection can be achieved through the use of protective equipment such as edge pads (Fig. 7.2) and edge rollers (Fig. 7.3).

Figure 7.1 A rope bends over an edge.

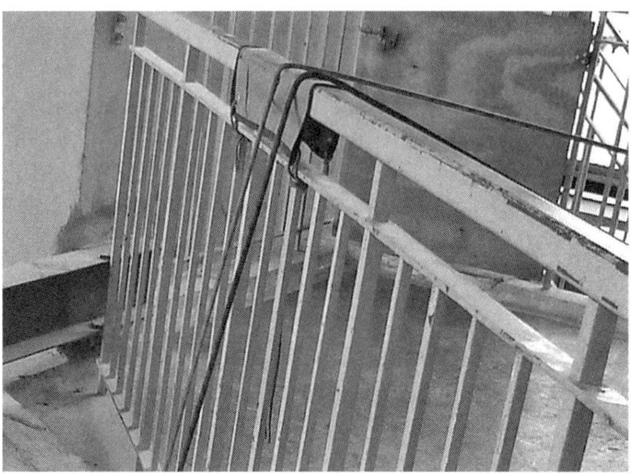

Figure 7.2 Canvas edge pad.

Figure 7.3 Edge roller.

Figure 7.4 Low-friction edge pad.

Either edge pads or rollers may be used for protecting stationary ropes, but edge rollers are often the better choice for moving ropes. Low-friction edge pads made of plastic (Fig. 7.4) or other smooth materials, though not quite as efficient as a roller, can also be used for moving ropes.

Friction isn't all bad, though. Some equipment is designed expressly for the purpose of imparting friction on the rope. Braking devices such as that shown in Figure 7.5 are an excellent example of this. The basic means of friction employed by braking devices is simply that of friction against the surface of the device combined with angles made by the rope as it is reeved through the device.

The more variable the friction imparted by a descender, the greater the range of control offered by the device. Many modern descenders offer variable friction through a mechanical means, such as a cam device, a bollard, or some other movable component. A brake rack is a common example of a device with variable friction. See Chapters 4 and 5 for further discussion on descenders for personal and rescue

Figure 7.5 Friction-based descender.

use, respectively. Regardless of how friction is applied, it inevitably creates at least some compromise in strength at the point(s) where it contacts the rope.

Managing the effects of gravity and friction, together, is foundationally critical to most technical rescue operations. Meanwhile, another important principle comes into play: that of *angles*.

ANGLES

Rigging angles are an extremely important consideration in rescue. Angles occur wherever a rope changes direction in a system, which occurs at the transition point where a rope goes over an edge, almost anywhere that a piece of auxiliary equipment (such as a pulley or connector) is used in conjunction with the rope, and in knots and bends used to terminate a rope. Angles are a consideration in anchor rigging, horizontally rigged systems, vertically rigged systems, essentially anywhere a rope is used. As in the case of friction, properly managed angles can offer distinct advantage and usefulness, while unplanned-for angles can be a detriment to the system.

Wherever a rope incurs a bend at least some loss in strength of the rope at that point must be presumed. How much loss depends on the nature of the bend, and whether it is around a tight or a narrow pivot point. Strength loss of a rope around a bend begins to become quite significant at a *d:D ratio* of less than 1, that is, any time the bend in the rope is tighter than 1× the diameter of the rope. The tighter the bend, the greater the strength loss. We see this in the case of knots, and where a rope is bent over or around another object, such as a connector or a pulley. For modern kernmantle rope, this strength loss becomes negligible at about twice the diameter of the rope.

In addition to affecting strength of the rope, a bend will also affect the efficiency (or energy loss) in the system. A system consisting of a rope that travels in a straight line without contacting any object sees no efficiency loss, whereas a system

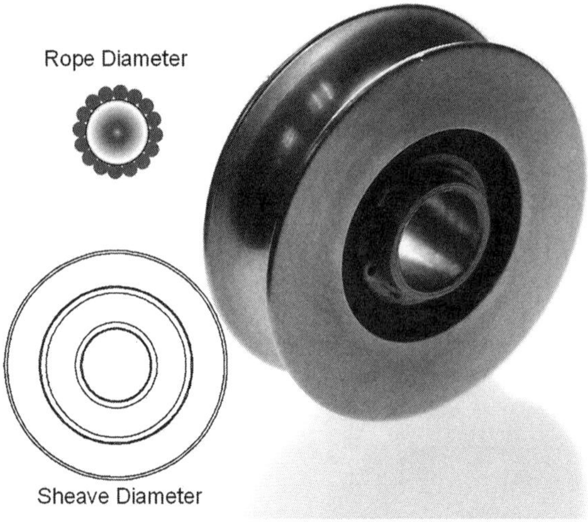

Figure 7.6 Pulley sheave diameter as compared with rope diameter.

consisting of a rope bent around a pulley will experience efficiency loss, even if the pulley has very efficient ball bearings. Compare the diameter of the pulley sheave with the diameter of the rope (Fig. 7.6). Again, the tighter the bend the greater the efficiency loss. All other things being equal, at a radius of about 4× (or greater) the radius of the rope, efficiency loss as a result of the bend becomes insignificant.

Vector Forces

A word of caution is in order regarding angles in systems. In a straight line system there are only two forces acting on the system: the load, and the anchor. When a pulley is placed into a system, a third force is introduced by the angle at which the rope travels through the pulley. As the load accelerates, the resulting forces on various points in the system will vary. You need not be a physicist, but you should understand this concept of force magnification.

First, let's consider the effect of the load. If a load weighs 100 kilograms and is equally supported by two "legs" of a single rope, as long as the two legs of the rope are more or less parallel, each line supports only about half the combined load and the force on the anchor is no greater than the force of the load itself (Fig. 7.7).

In reality, however, we would seldom bend a rope around a pulley and then connect both ends of that rope to the load. The reason for a pulley is usually to facilitate movement, so normally we would connect *one* end of the rope to the load, and we would *pull* on the other end. Again, assuming that the load weighs 100 kg and is connected to only one leg of the rope, we must exert an equal force (100 kg) on the other leg (e.g., the working end) in order to keep it from falling. With relatively parallel legs, as in Figure 7.8, the force at the anchor is equivalent to the combined load on the two legs, in this case, 200 kg force.

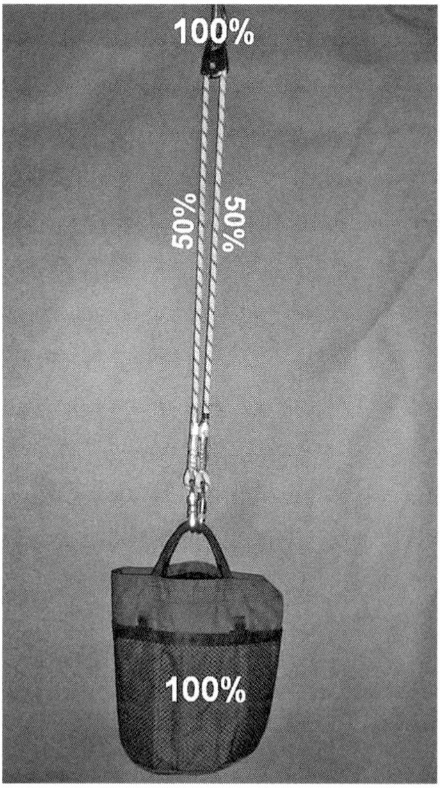

Figure 7.7 A load equally supported by two “legs” of rope around a pulley.

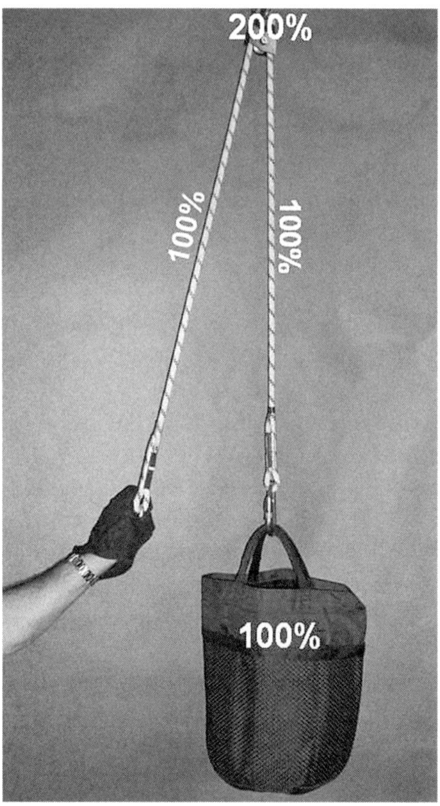

Figure 7.8 Force at the anchor is equal to the combined loads on the legs as long as the lines are nearly parallel.

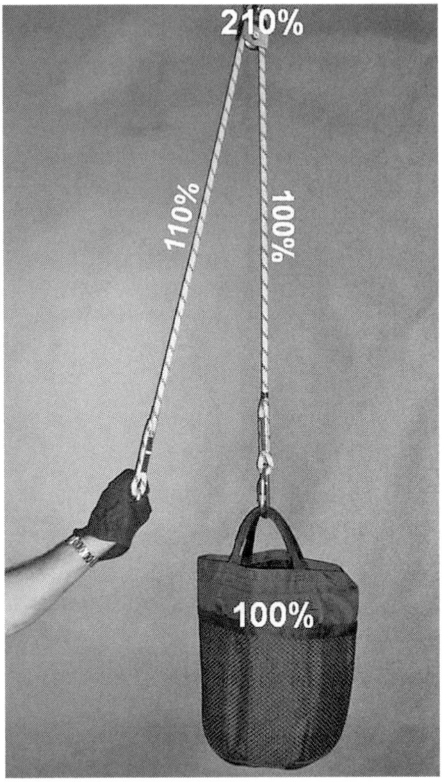

Figure 7.9 Additional force must be exerted on the working end to move the load. Force at the anchor remains roughly equivalent to the combined total.

When we want to move the load, we must exert additional force on the working end in order to overcome inertia and move the load. Still, the anchor will see roughly the equivalent of the combined load on the two legs. So, if we have a 100 kg load, and we exert 110 kg force in order to overcome inertia to move that load (Fig. 7.9), the force at the anchor will be about 210 kg.[2]

For illustration purposes, for a moment let's go back to considering a static (unmoving) load as illustrated in Figure 7.8. In that example, the two legs of the rope are roughly parallel as they exit either side of the pulley. However, it is notable that as the interior angle of the bend increases (becomes wider), the forces on the anchor, pulley, or bollard at the bending point can increase significantly.

The magnitude of a force as it is magnified by the imposition of an angle is known as a vector force. This increase becomes even more significant when a load is in motion. A firm understanding of vector forces becomes especially important in rigging artificial high directionals, haul systems, and mid-line anchor points to re-direct a rope.

[2]This example is approximate, for illustration purposes, and does not take into account frictional losses.

EXERCISING THE CONCEPTS

When a worker falls and is either suspended from his or her fall arrest system or has landed on a lower level, rope systems may be devised to raise or lower the subject to a more secure location, as necessary. A seriously injured person should be moved only by properly trained emergency/medical responders, except to remove him or her from immediate danger or to render immediate life saving assistance. Professional emergency response is beyond the scope of this text; here we will address only co-worker assisted rescue, that is, immediate assistance that might be rendered to an uninjured person, or to one who is in imminent danger and must be moved.

Rope rescue can generally be classified into two broad types of operations: *lowering* and *raising*. Both lowering and raising operations can take place on low to high angle terrain, including vertical and free hanging situations. As previously discussed, friction is the primary means of limiting or preventing a loaded rope's ability to travel. In both high and low angle lowering systems, friction is used to control the speed of descent of the load, whether it consists of a patient, rescuer(s), litter, or some combination thereof.

There are scores of training organizations committed to teaching the subject of rescue but each of these, it seems, strives to be unique in terms of equipment, techniques, and methods of accomplishing the task. While this may appear to be a potential source of controversy, in fact it illustrates perfectly that the number of effective and appropriate ways to execute any given task within a rope rescue is virtually unlimited.

While the basic elements of rope rescue—gravity, friction, and angles—remain present regardless of the environment, selection of tools and techniques will vary based upon local preference, compatibility of components, regulatory requirements, training, experience, and scene considerations. The co-worker assisted rescuer is challenged to see these variables as opportunities to expand his or her knowledge and exercise his or her skills, all the while maintaining and prioritizing safety.

Clearly there are advantages in an employer selecting a single system and method as a preferred solution for rescue for his or her workplace. However, consistency should not become a synonym for complacency. The advantage of versatility should not be overlooked. Whether using a *pre-engineered rescue system* or using components to create systems in the field, understanding the fundamental concepts behind rope rescue will help the responder to apply them with fresh perspective to each unique situation.

Such a conceptual approach begins with considering and prioritizing the fundamentals of what you are trying to achieve, and then determining whether to achieve these by way of equipment, techniques, or some combination of the two.

Anchor Systems

One of the most complex problems in any rope rescue is the anchorage. In the event of an incident, Murphy's law dictates that anchorages are unlikely to exist at the place and time where they are needed in an emergency. Especially in urban, industrial, and confined space environments, anchorages are often unavailable, or if they do exist, are questionable.

All rope systems must incorporate solid, well-constructed anchorages as a foundational element. The anchorage is typically a fixed structural member that offers sufficient stability and strength to serve as the termination point of the system, and to hold the system in place. While it is possible to perform rescue using an anchorage that is on the same level as the subject, selecting a high anchor will make the process easier.

Anchorages for use in emergency/rescue situations should be selected, identified, and marked as part of the pre-planning process. Predicting emergencies is not an easy task, and predicting where, exactly, a subject might need to be lowered or raised can be even more difficult. Nevertheless, this is an important part of the rescue preplan.

Anchor systems are a rather hotly debated and somewhat confusing topic, largely as a result of inconsistencies in terminology. Establishing a good baseline of accurate terms is key to communication between members of a workplace response group.

- Anchor Point (see also Anchorage): A place, fixing, or fixture that supports and to which the various ropes and rope systems are attached.
- Anchorage (see also Anchor Point): A place or fixture that supports and to which the various ropes and rope systems are attached.
- Anchorage Connector: The means of securing to an anchorage, such as a strap or eye bolt.
- Anchor (or Anchorage) System: The combined anchorage and anchorage connector, working together to support a load.

It is the term "anchor," when used alone as a noun, that can be most misleading, and the confusion is usually over whether the term might mean a mere anchor point or a fully fledged anchor system. In order to avoid this kind of confusion, it is best to simply avoid using the word "anchor" as a stand-alone noun. The term "anchor" may appropriately be used as a verb in common rigging vernacular, that is, "to anchor the rope," and indeed you may find it used in this manner in this book.

Anchorages

Anchorages are chiefly found in one of two forms: structural or installed. Structural anchorages generally consist of existing structures (Fig. 7.10)—for example, I-beams, human-made structures, healthy trees of good size and mass, large boulders, and heavy equipment. These items are perhaps not specifically designed to be used as anchorages, though some human-made versions may have special attachment points welded or bolted into place. Generally, structural anchorages are an excellent choice for ropework, but care must be taken to verify the security and appropriateness of the structure for such use.

Installed anchorages are those that are placed expressly for the purpose of being an anchor. A davit arm on top of a building, a removable bolt, and an eye-

Figure 7.10 Any structural element may be considered an anchorage.

Figure 7.11 An installed anchorage.

bolt (Fig. 7.11) are all examples of installed anchorages. Especially in the case of permanently installed anchorages, the lines between what constitutes an anchorage as compared with an anchorage connector can sometimes blur. Philosophical discussions aside, what really matters is the performance of the anchorage system as a whole. Installed anchorages may be installed so as to be permanently fixed, or they may be temporary in nature. Pre-planned and pre-installed fixed anchorages are advisable in locations where work at height or rescue is likely. Such an anchor is normally referred to as an *engineered anchor*, and in some states it is required that a professional engineer perform the installation and/or verification of the anchor. In such a case, this is called a *certified anchor.*

In establishing or selecting an anchorage, the user must take into consideration the forces likely to be generated by reasonably foreseeable loading, potential direction and magnitude of loading, deflection of the anchorage, and the impact on the structural members to which the system is attached.

Anchorage System Performance

Certified anchorages for fall arrest should be rated to at least two times the maximum potential arrest force (in event of a fall), multiplied by the number of authorized connections. Only a Qualified Person can establish a certified anchorage. ANSI Z359 does an excellent job of identifying appropriate anchorage strength requirements for non-certified anchorages, as outlined in Table 7.1.

Where structural anchorages are limited in availability, several temporary anchorage alternatives are available. Tripods (Fig. 7.12) are an excellent anchorage solution for many confined space and other applications.

When considering a tripod, important factors include number and type of head attachments, tripod height and strength, and optional accessories. The unique SMC TerrAdaptor (Fig. 7.13) is an ingenious twist on the original tripod concept that also works well over an edge.

This modular unit (multi-pod) consists of a series of versatile leg components and connecting elements, allowing it to function as a tripod, an A-frame, a gin-pole, or even a quad-pod. Available with interchangeable foot configurations, its adjustable design is adaptable to any terrain, enabling rescuers to establish a high point connection in almost any location. It can be used directly as an anchorage, as a change of direction, or simply as a high point for lifting a rope system off the ground.

For urban applications where a tripod is not feasible, davit arms (also sometimes called "needles") can be useful. These are generally "borrowed" from other industries such as swing-stage suspension, and co-worker assisted responders should use these only if properly trained and then only within the scope of their intended use. Davit arms may be permanently mounted with a fixed base, or may be used with a portable base. The strength of a davit arm is generally quite limited, with a design load of around 350 lbs. Especially when used in a portable configuration, the

TABLE 7.1 Appropriate Anchorage Strength Requirements for Non-Certified Anchorages

Intended Use	Minimum Strength	Multiply by
Restraint	1,000 lbs (4.5 kN)	Number of authorized connections
Positioning	3,000 lbs (13.3 kN)	Number of authorized connections
Rope access	5,000 lbs (22.2 kN)	Number of authorized connections
Vertical lifeline	5,000 lbs (22.2 kN)	Number of authorized connections
Rescue	3,000 lbs (13.3 kN)	Number of authorized connections

Figure 7.12 Typical tripod. Credit: John Peleaux, Innovative Access, Inc.

Figure 7.13 TerrAdaptor offers unique rigging adaptations.

Figure 7.14 ClimbTech I-beamer. Credit: ClimbTech.

unit must be counterweighted and/or back tied and secured. Davit arms are occasionally found on the tops of buildings, where they are used to support suspended scaffolds or maintenance workers on rope. These should not be used in an emergency situation without first verifying the appropriate use, limitations, and application of the specific unit that is being used.

Beam clamps are another type of versatile anchorage connector that are especially useful in industrial environments. These are designed specifically to be mounted on I-beams and similar structures where anchorages can be difficult to establish. Figure 7.14 shows one example of a versatile beam clamp.

Rescue anchorages should be selected to be as high as practicable above the system to be used, in order to minimize potential fall distance and to facilitate ease of working with the system. When an anchor itself is not of sufficient height, or where a high anchorage is not available as near to the edge as the rescuer might wish, the anchor may be established far back from the edge and the line passed through a high point at or near the edge. This type of setup is known as an "artificial high directional" or simply "high directional" (Fig. 7.15).

Great care should be taken to ensure that additional hazards are not introduced by angles that are created as part of the high directional. The rigging and angles should be such that they impart appropriate compressive forces on the system, and that the equipment used is able to withstand these forces. The effects of vector forces, as discussed earlier in this chapter, should be managed in terms of both actual force and direction of pull. Care should especially be taken so that the direction of pull does not create a tipping action on the artificial high directional equipment used. Special training is required for the use of this equipment.

Co-worker assisted rescue systems are unique in that anchorages may have to sustain a two person load (rescuer and victim), and may also be exposed to limited dynamic loading. Although the rescue methods selected should seek to eliminate dynamic loads, potential for both static and dynamic forces should be taken into consideration when considering rescue anchorages. Where a safety line is used in conjunction with the main line, each line should have its own separate anchor system and should be separately fixed to the worker's harness. This does not preclude both lines being attached to a single harness attachment point. This dual connection concept is based on the premise that simultaneous loading of both lines is highly unlikely; loading of the safety line is likely to occur only in the event of a main line failure.

Any rescue anchorage should be capable of withstanding force in any direction and to any magnitude that may be permitted by the anticipated rescue system.

Figure 7.15 TerrAdaptor as artificial high directional. Credit: Pigeon Mountain Industries, Inc.

Direction of pull is much easier to predict for fall arrest anchorages than for rescue anchorages. One method of predicting direction of pull involves following the fall line. The fall line can be described as the direction a ball would take if dropped or rolled from a given point. This, too, can also be assumed to be the direction that a rescue load would take from the point of the anchorage, given the pull of gravity.

The strength of an anchor system is completely dependent upon the strength and configuration of the components that connect the system to the anchorage. The connecting components should be of sufficient strength for the task to be performed, and should be attached in such a way so as to prevent unwanted movement or disengagement of the rescue system from the anchorage.

Evaluation of anchor systems is best done by experienced personnel, as this is a somewhat subjective process and requires training and experience. Before any anchor system is subjected to a live load, a safety check should be made to ensure that:

- The anchor system is able to withstand the load.
- The overall strength of the anchorage is sufficient and offers an adequate margin of safety.
- The anchorage is secure, and that it is not likely to slip or move when loaded.
- The anchorage and anchorage connector(s) are in good condition.

All potential anchors must be thoroughly evaluated by a Competent Person before use in order to ensure that overall system safety factors can be met. If sufficient connection exists directly to the anchorage, this is the preferred approach. Otherwise an additional component may be utilized to offer interface to the anchorage (e.g., bolt, D-ring, strap, etc.). Such components are generally referred to as anchorage connectors. Together, the anchorage and the means of connecting to the anchorage make up the anchor system.

Rollout

Compatibility between the anchorage and anchorage connectors is especially important, given that most anchorages are fixed in position during use. Some of the geometries achievable between an anchorage connection and anchorage can result in a magnification of the applied load on portions of the anchorage connector and/or anchorage. Incompatibility between anchorages and anchorage connectors can also contribute to rollout, where connecting components press against one another in such a way so as to cause one or the other to become dislodged. Rollout can also occur if the system is pulled in a direction that might not have been anticipated. This concept was introduced in Chapter 5 in terms of equipment, but it bears repeating here, in the context of rigging. After all, it is not the equipment itself that creates a rollout condition, but rather how the equipment is rigged and used.

For example, eye-bolts and D-rings are quite common anchorage connectors. The simplest method of connecting to this type of anchor is to clip a carabiner or a snap hook into the D-ring or eye bolt. However, the size and shape of most conventional D-rings and eye bolts is such that a carabiner or snap hook can easily be twisted in such a manner that inward or outward force is exerted across the gate. In this case, the pivot point can allow a force of up to two times the load to be imparted directly on the gate, which is usually the weakest part of the connector (Fig. 7.16).

Similar effects can occur any time multiple hard links are placed in line with one another. The less flexibility there is in a system, the more likely it is that a rollout condition might be created.

Of course, the best solution is to rig and use systems in a way that prevents gates of connectors being subjected to rollout, but this may not always be feasible. Careful selection of components can go a long way toward preventing this kind of situation. Consideration of components in relation to how they will be utilized and ensuring that the equipment will work properly together is essential to the concept of *compatibility*. Just ensuring that two components meet a given standard does not necessarily ensure compatibility between those two components. A Competent Person must carefully consider the geometry of components, how the equipment is being used, and potential conflicts between those components.

If utilization of non-compatible anchorage connectors is unavoidable, one method of avoiding rollout is to insert a *soft link* into the system (Fig. 7.17). A soft link is simply a webbing or rope loop that acts as a flexible member in a system to prevent rollout.

Figure 7.16 Avoid rollout conditions.

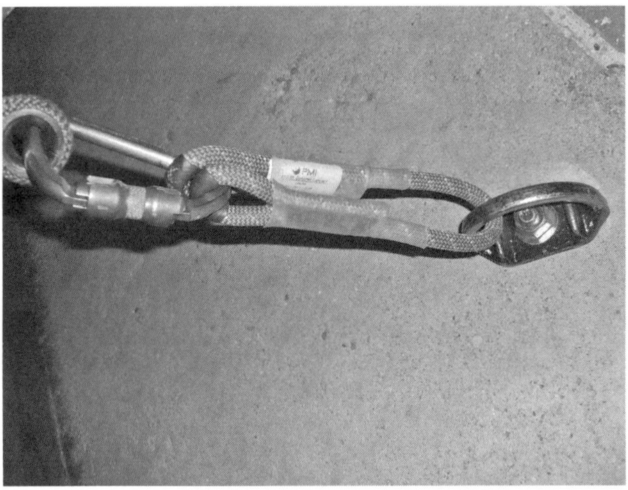

Figure 7.17 Use of a soft link to help prevent rollout.

Tensionless Anchorage

Another method of overcoming the issue of non-compatibility is by using *tensionless anchors*. A tensionless anchor (Fig. 7.18) is a minimalist approach to securing an anchorage. Only the anchorage and a rope are required, and the security of the anchorage relies on the principle of friction. The best anchorage for a tensionless anchor is a fixed structure whose edges are not too harsh. It may be padded as needed to protect the rope or the structure. The optimum anchorage diameter is at least four

Figure 7.18 Tensionless anchor.

times the diameter of your rope, but not so great a diameter that it wastes valuable rope length.

To create a tensionless anchor, simply wrap the standing end of the rope around the anchorage several times. In practice, three to four wraps normally results in enough friction that pulling on the working end of the rope will not result in movement, but the diameter of the anchorage and the coefficient of friction will also have an effect. For an added measure of security, knot the standing end of the rope, clip a carabiner into it, and then clip the carabiner loosely over the working line as it exits the tensionless anchor (Fig. 7.19). Take care when clipping off the standing end of a tensionless anchor to avoid creating a change of direction in the working line, as illustrated in Figure 7.20.

Load-Sharing Anchor Systems

A load-sharing anchorage, also known as a load-distributing anchorage, is a type of anchorage that shares (or distributes) the load between two or more anchor points. The anchor points are usually positioned somewhat adjacent to one another and connection between them is generally achieved by using some variation on a looped rope, cord, or length of webbing. Load-sharing anchorage systems may be useful when the anchorage does not exist in the best possible position, or where it might be advantageous to reposition the anchor.

In the simplest form of a load sharing anchor, a single sling is placed into two anchor points, and a knot is used to form a loop between the two anchors, positioning the load at the tangent end of the anchor. Using this method, the location of the connecting point can be adjusted to accommodate a variety of fall lines.

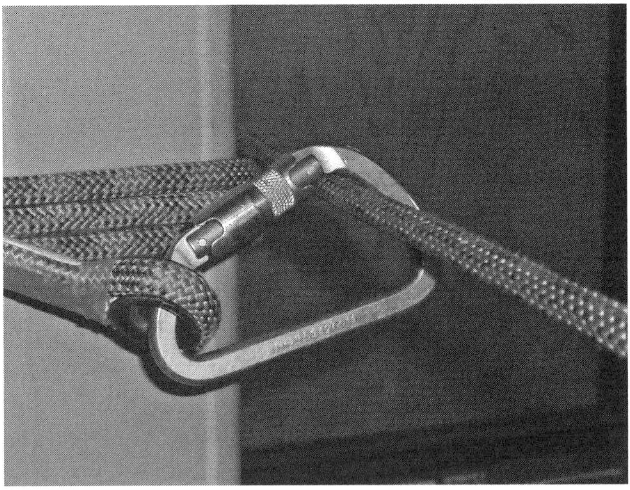

Figure 7.19 Clip the standing end of the tensionless anchor loosely onto the mainline.

Figure 7.20 Take care to not create a change of direction when clipping off the standing end of a tensionless anchor.

A similar result can be achieved by using two separate slings, one to each anchor point, and bringing them together to clip into the load line as shown in Figure 7.21. A load-sharing anchorage is typically considered one anchorage, even though it may be constructed of two or more anchor points. This is because the integrity of the anchorage as a whole relies on the interdependence between the anchor points. One load-sharing anchorage system equals one anchorage for one line. If a decision is made to use two systems, a primary and a secondary, separate anchorage systems should be constructed.

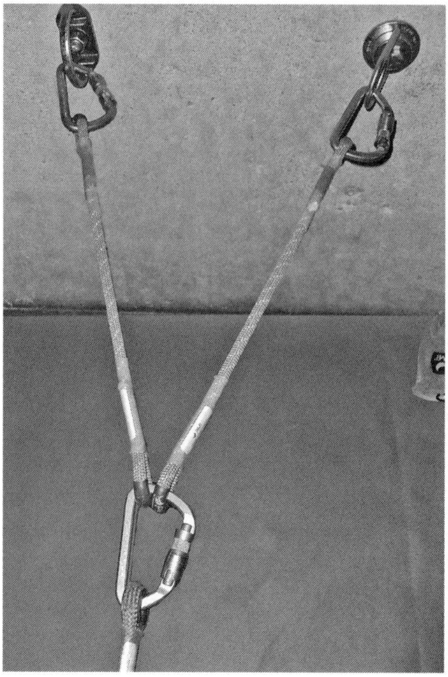

Figure 7.21 Two-point load-sharing anchor, two slings.

One potential disadvantage of a fixed type of load-sharing anchors is that as the direction of pull changes (e.g., if the fall line varies or the rescuer must deviate to one side or the other), either one or the other anchor might be subjected to the entire load in a direction that may not have been intended. This somewhat defeats the whole concept behind the load-sharing anchor, as one or the other of the anchor points may in fact be subjected to the entire load at various times.

Such disproportionate loading between anchor points may be avoided through implementation of techniques that permit the system to self-adjust. In the self-adjusting version of a load sharing anchor, the load remains distributed more or less equitably between the anchor points, even as the load moves across the fall line. To create a self-adjusting load sharing anchor system (Fig. 7.22), follow these steps:

1. Select two or more appropriate anchor points.
2. Clip a sling between the two anchor points.
3. Make a twist in the line and clip into the X, as shown.

While the load-sharing anchor system should be built so as to avoid any potential of failure in any of the legs, consideration must be given to the consequences of a failure. The purpose of the twist introduced in Step 3 is to ensure that if one anchor point should fail, the remaining leg is "captured." Simply clipping into the bight formed between the two legs, without the twist, could permit catastrophic failure in the event that one anchor point were to fail.

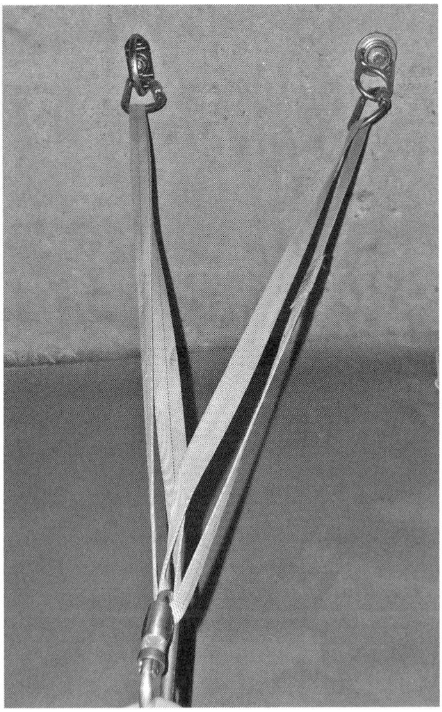

Figure 7.22 Self-adjusting load-sharing anchor—two point.

Potential impact force is also a consideration. If too long a sling is used to connect the anchor points, the fall could potentially be a long one that would result in high impact loads. This potential hazard can be avoided by extending each anchor point with an anchor strap, as in Figure 7.23. Extending the length of your anchorage connectors can help reduce the distance and force that may be generated if one anchor point fails. These can then be connected with a very short sling to provide the equalization function. A sling made of a slippery material such as high modulus polyethylene (HMPE) works well for this purpose. Although HMPE has little to no force absorption capability, the slippery nature of these fibers allows for better equalization of the load, and the shorter sling helps to prevent a fall of significant distance.

One disadvantage of extending anchor points as shown above is that it requires the use of additional components, thereby increasing the complexity of the system. This potential disadvantage must be weighed against potential advantages, and a cost/benefit decision made based on circumstances. There is no one perfect solution for all situations; rigging requires constant vigilance, and the "best" choice for a given situation made based on consideration of all factors.

Whenever multiple anchor points are used together, whether in a static or self adjusting version of a load sharing anchorage system, the angle formed between the legs of the anchor materials should be as low as possible so as to reduce the potential forces on the respective anchor points. This is because the amount of force that is

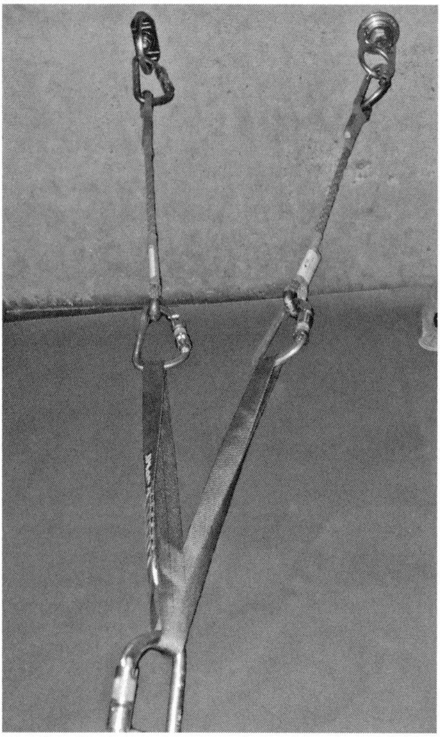

Figure 7.23 Extend your anchorage points to prevent significant fall distances.

applied to each anchor point increases as the angle increases. As discussed earlier in this chapter, angles affect the force that is applied in a system. In a two-point anchor system with a 120° angle between the legs, each leg sees 100% of the load, as illustrated in Figure 7.24.

As the angle exceeds 120°, magnification of forces on the two anchors becomes even more extreme. By the time the angle is at 175°, each anchor can be exposed up to 11 times the force of the load. For this reason, some agencies suggest up to a 90° angle as a rule of thumb and 120° as an absolute maximum, referring to this as the "critical angle."

The hazard that is created in establishing a rule such as this is that responders may come to rely on the rule in lieu of really understanding the effects of the angles on the system. It is most important that responders understand how forces may be amplified, and that the anchorages and other parts of the system are able to withstand whatever forces rescuers impose (plus an appropriate factor of safety). In other words, even 120° may be too great an angle for your system, if your anchorages and components cannot all withstand 100% of the load (plus safety factor). On the other hand it may be perfectly fine to exceed a 120° angle in an anchor system as long as your anchorages and components can all withstand several times the force of the load (plus safety factor).

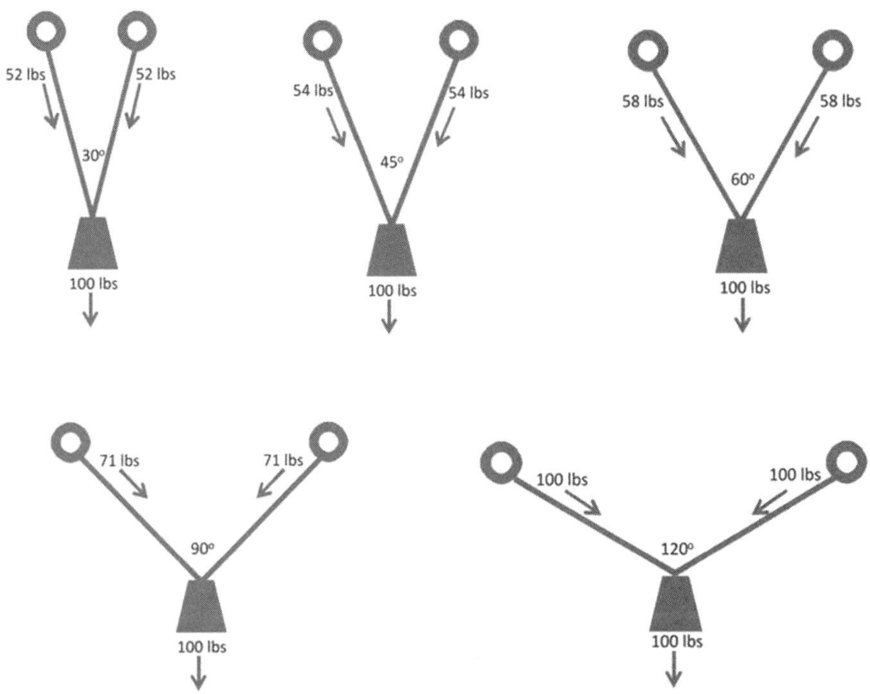

Figure 7.24 As angle increases, so does force on each leg of a two point anchor.

Figure 7.25 Anchor with back tie for stability.

Back Ties

A *back tie* (Fig. 7.25) may be incorporated into the anchor system where extra security is deemed desirable, or where there is concern that an anchor may shift or creep during use. A back tie is not in itself considered to be an anchorage, and it need not necessarily be of equivalent strength to the primary anchorage. The purpose of a back tie is simply to provide security and stability to an anchorage.

The intricacies of back ties are an advanced skill, and are beyond the scope of this book. The concept is noted here only so that the reader is aware and will recognize the concept if it is encountered in the field.

Change of Direction

When it is anticipated that the rescue system or load might need to deviate from the fall line or the anticipated direction of pull, it may be appropriate to establish an additional anchor to create a change of direction.

There are two quite different approaches to changing direction. The first, called a *directional deviation*, is simply a midline anchor into which a line is clipped but not independently tied off.

In a directional deviation, the line is allowed to run freely through a connector or pulley device. The performance requirements of an anchor system for a directional deviation will depend on the intended load, and the probability/consequence of failure. Vector force amplification should be taken into consideration, as this may necessitate strengths even higher than the original anchorage requirements. The force at a directional anchor can reach 1.4 times the force of the load with a 90° bend in the line, and double at 180°. Care also must be taken when using this type of system to avoid potentially dangerous pendulum falls.

Another approach to creating a change of direction is to re-anchor a rope mid-line, so that the length of rope above is essentially independent from the rope below. This method is called a *rebelay*, or *re-anchor*. When re-anchoring a rope, sufficient slack should be left in the top length of rope to ensure that the upper section of rope will not become trapped or encumbered by the re-anchor during use.

A re-anchor system must be at least as strong as the original anchor, or five times the potential load—whichever is greater. Considering previous discussions regarding the effect of angles in a system, calculating potential load on a directional deviation can be particularly challenging.

RESCUE SYSTEMS FOR LOWERING

When it becomes necessary for one person to move another to safety, lowering the subject is often a good choice. One benefit of using a lowering system for rescue is that the rescuer is able to use gravity to his or her advantage, reducing the amount of exertion required.

A lowering system requires at least the following parts, shown in Figure 7.26:

1. Anchorage
2. Rope
3. Braking device

and, of course, the necessary connectors to hold it all together.

The anchorage for a lowering system should be positioned above, in line with, but not creating a hazard to, the subject below. Appropriate connection should be

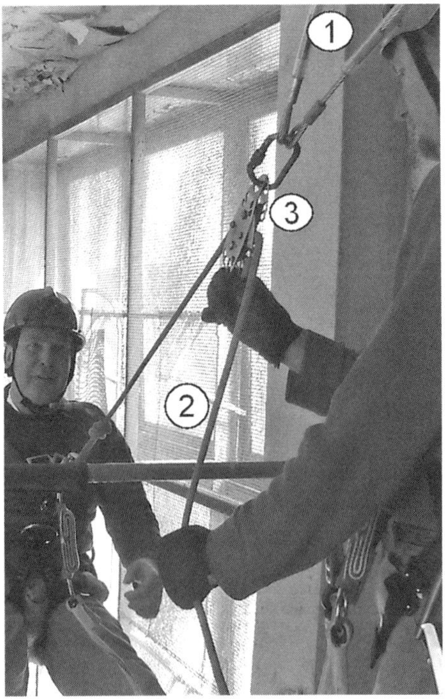

Figure 7.26 Parts of a lowering system.

made to the anchorage by an anchorage connector. Of course, the details of the connection will vary depending on the anchorage selected and the equipment available.

A braking device (see Chapter 4) is then attached to the anchorage connector, the rope is passed through the braking device, and finally, a connector on the running end of the rope allows the load/subject to be attached. Specific systems and methods for lowering are discussed in greater detail in Chapters 8 and 9.

RESCUE SYSTEMS FOR RAISING

Rescuing a subject sometimes requires that the subject be raised to a higher level. The easiest way to accomplish this is with a winch (Fig. 7.27) or similar mechanical device. Such devices should always be designed to be fail safe and should only be utilized if they are specifically designed and approved by the manufacturer to support human loads. Most winches have the capability to both lower and raise. This is an important consideration, especially in rescue. Whenever a rescue load is being moved in one direction, provision should also be made for moving the load in the opposite direction in case of a problem. For example, consider if a subject who is working in the steel structure beneath a bridge were to fall unconscious. Co-workers may use a winch to raise him to the deck, but the many angles and connections in a bridge structure offer ample opportunity for an incapacitated subject to become

Figure 7.27 Rescue winch. Credit: Donald Enos/SMC.

wedged in a joint. If the winch is equipped with a kill switch, the increased force on the winch would cause the kill switch to trip. The subject could then be lowered slightly and dislodged from the jam before continuing to raise. If the winch is not capable of immediate action in both directions, such an incident could be catastrophic.

Hoists and winches are a good choice for workplace application, but just having a winch on site is not necessarily an appropriate "easy answer" to all rescue needs. A winch must be rigged and used properly, just as any other equipment. Because many winches can be operated through the moving of a lever, turn of a wheel, or the press of a button, the importance of proper training is easily overlooked. Training is required for any rescue activity.

> A fail safe mechanical winch design is a one that will not freewheel or otherwise drop the load if the raising mechanism should accidentally fail. A winch used for rescue should also have reverse capability.

Winch/hoist systems are often pre-rigged at locations where there is a known high probability that they will be needed. A disadvantage of mechanical raising devices is that they are all too easily relied upon, but are not always where you need them to be when you need them. They can also be quite expensive, heavy, and require special brackets for connecting to an anchorage.

Would-be rescuers should also be familiar with manual forms of raising. Manual raising methods are easily transportable, do not require AC or DC power, and can be adapted to multiple environments. At the most basic level, a raise might simply consist of a rope with a subject (or load) at the bottom end and a rescuer pulling on the top end. Obvious hazards notwithstanding, let's explore the mechanics of this type of raising system.

In this direct pull method (Fig. 7.28), the force pulling the rope must be equal to or greater than the force of the load to overcome inertia and start the load moving;

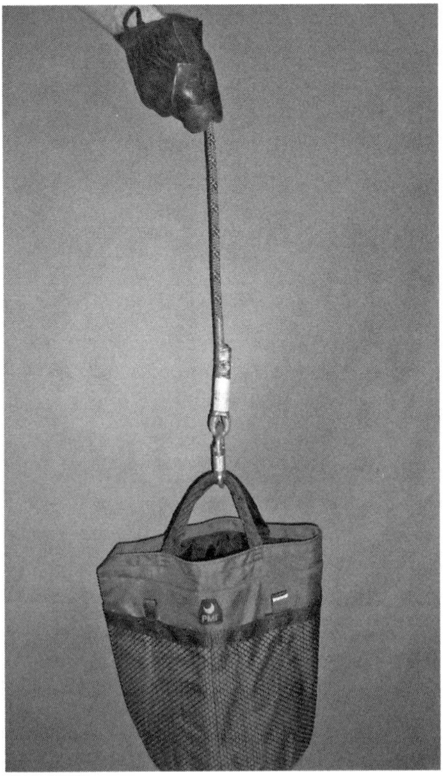

Figure 7.28 Raising with a direct pull.

in other words, the amount of force required to move the load represents a 1:1 relationship with the load itself. This 1:1 relationship can be seen by the fact that in order to lift a 200 lb load, the rescuer would have to use at least 200 lb of force (plus a little extra to overcome inertia and system friction).

We can express the simple example shown in Figure 7.28 as a ratio, as follows:

Weight of load / Force it takes to move load

200 lb / 200 lb

1 / 1

So the ratio is 1:1. Notice that in this example the length of rope pulled by the rescuers is the same as the distance the load moves, also reflecting a 1:1 ratio.

The direct-pull method may be a viable approach (additional safety measures notwithstanding), but it relies heavily on the strength of the rescuer in relationship to the weight of the subject. Practically speaking, it would be very difficult for a 200 lb rescuer to pull a 200 lb subject even as little as 10 feet in this manner.

This is where *mechanical advantage (MA)* comes in. The amount of force that a rescuer must exert in order to move a load can be greatly reduced by using pulleys and rope to create a mechanical advantage. This type of system is commonly referred to as a *haul system*. With a mechanical advantage haul system, a relatively heavy load can be moved with relatively little force. Depending on the types and configuration of pulleys and rope, varying amounts of mechanical advantage can be produced to reduce the amount of force a rescuer must use to move a load.

The amount of mechanical advantage produced by a given system is calculated by formulas and is expressed as a ratio. For example, if a rescuer builds a system wherein a 200 lb load could be moved with a calculated 100 lbs of force, he would call this a 2:1 mechanical advantage. This calculated mechanical advantage is referred to as *theoretical mechanical advantage (TMA)*.

The reason that we refer to the calculated mechanical advantage as theoretical is because in reality some of the energy is lost through things like friction in the pulleys, rope rubbing or dragging against an edge, and rope stretch. The resulting mechanical advantage, after friction and other losses, is known as the *actual mechanical advantage (AMA)*.

For discussion purposes in this text, we will discuss hauling systems in terms of TMA.

> The actual amount of MA you do get out of a hauling system will depend in great part on how efficiently you can rig the system and on the efficiency of the components such as pulleys you use to build it. Direction of pull, quality of components, and limiting friction in the system will all help improve your AMA.

To reduce the amount of force that the rescuer must exert in order to move the load, the mechanical advantage can be increased simply by the introduction of a single pulley, placed at the load itself. A pulley in this position is known as a traveling pulley because it travels with the load. With one end of the rope anchored at the top of the drop and passed through the pulley at the load (Fig. 7.29) the rescuer(s) pulling upward on the other end of the rope realize a 2:1 theoretical mechanical advantage. That is, with the 200 lb load supported by two strands of rope, one attached to the anchor and the other hauled by the rescuer(s), the rescuers feel only half the load, or 100 lbs.

Expressing the load information in this second example as a ratio looks like this:

Weight of load/ Force it takes to move load

200 lb / 100 lb

2 / 1

So the ratio is 2:1. This ratio, then, becomes the means by which we express the mechanical advantage in the system. It is said to be a 2:1 mechanical advantage.

One disadvantage of this setup is that the rescuer(s) must now pull 20 feet of rope to move the load 10 feet—thereby exposing that full length of rope to friction, which reduces the AMA. Another disadvantage is that the direction of pull (e.g., upward) may not be the most ergonomically efficient approach.

Figure 7.29 A 2:1 mechanical advantage.

Changing Direction A second pulley added to this system at the anchor to re-direct the rope (Fig. 7.30) will not offer additional mechanical advantage, but it may make the direction of pull more convenient for the rescuer(s).

In this example, there are still only two lines directly supporting the load, and moving the load 10 feet will still require 20 feet of rope to pass through the mechanical advantage system (plus a little extra for the direction change, but we don't count that). In theory it still takes 100 lb of force to move the 200 lb load. So, the calculation is still the same as in the previous example, with a TMA of 2:1. The pulley at the anchor is stationary, merely offering a change of direction.

> To quickly determine the amount of mechanical advantage in a *simple mechanical advantage* system, count the number of lines actually supporting the load. As shown in Figure 7.22 only two of the lines actually support the load; therefore, it is a 2:1 mechanical advantage. The third line, because it provides only a change of direction, is not counted.

When the pulley is anchored and does not change position as the rope moves, we call it a change of direction, or a *fixed pulley*. The term "fixed" in this instance does not infer a pulley that has been repaired, but rather a pulley that is affixed in place. A fixed pulley does not offer mechanical advantage. Fixed pulleys can make

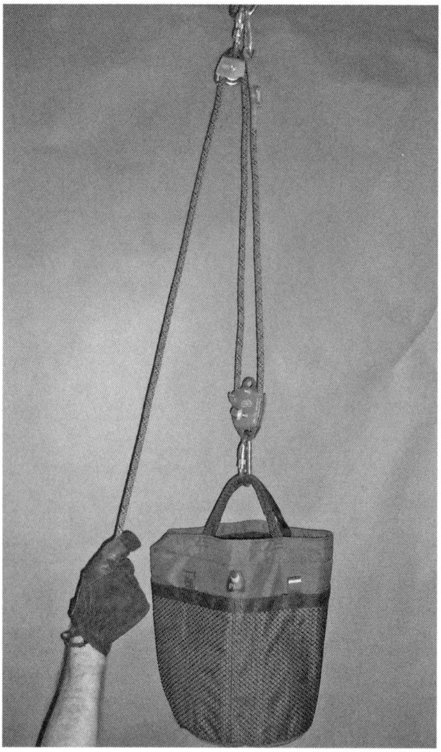

Figure 7.30 A 2:1 mechanical advantage with change of direction.

work easier by changing the direction of the applied force but the same force is needed to move the load. Use caution in these situations to not overload equipment and anchorages as a result of vector forces, discussed earlier in this chapter.

Carrying the concept one step further, if a third pulley is added back at the load, that pulley would move with the load—this is also known as a traveling pulley. Traveling pulleys *do* increase mechanical advantage, so in this case additional mechanical advantage would be introduced.

To determine the TMA in a system, count the number of lines that are supporting the load (Fig. 7.31). In this case, there are four lines supporting the load, ergo there is a 4:1 mechanical advantage. Moving our 200 lb load will require only 200 lbs/4, or 50 lbs, of force. A fourth pulley added to this system would once again provide only a change of direction, a fifth pulley would create additional TMA (6:1), and so on. Of course, at some point it becomes terribly inefficient to continue adding pulleys and in truth we seldom see more than a 4:1 mechanical advantage created in this manner.

Note that whenever the first pulley in a system is placed at the load, the result is an even numbered mechanical advantage. Whenever the first pulley in a system is placed at the anchor, the result is an odd numbered mechanical advantage.

Figure 7.31 Count the number of lines supporting the load.

Up to this point, you may have noticed that all of the mechanical advantage systems we have discussed are even numbered ratios (2:1, 4:1, 6:1). Odd numbered mechanical advantages are also possible, but to create these the rope must not be anchored at the top. Instead, the end of the rope must be tied off at the load, and the first pulley in the system will be placed at the anchor.

With one fixed pulley at the anchor, there is no mechanical advantage; there is only a change of direction (Fig. 7.32). Note that there is only one line directly supporting the load. Moving the load will require input force equal to at least 100% of the load. It is only by placing a second pulley in the system that a mechanical advantage is realized.

By counting the lines that directly support the load, you will see that this creates a 3:1 mechanical advantage (Fig. 7.33). For a given amount of force exerted on the system, you will realize three times that force at the load. So, to move our 200 lb load requires just over 66 lbs of force, and for every foot that the load moves, the rescuer(s) must pull 3 feet of rope.

The 3:1 mechanical advantage system, also called a "Z-rig" because the configuration of the system causes the rope to take on something of a Z shape, is quite commonly used in rescue systems because it provides a good balance between mechanical advantage offered, amount of rope used, and internal system friction.

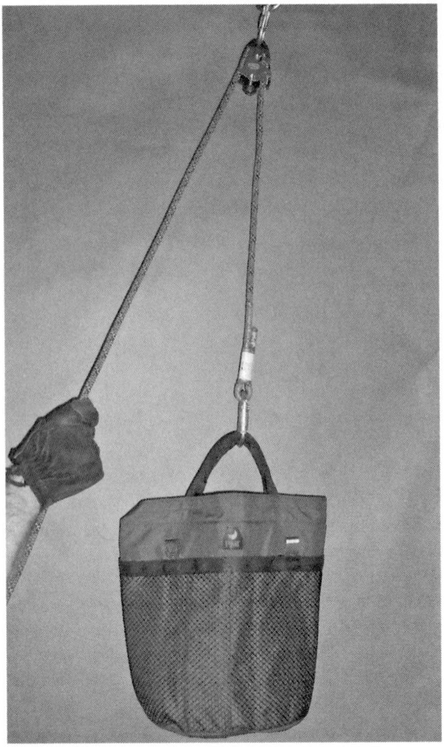

Figure 7.32 A 1:1 mechanical advantage with direction change.

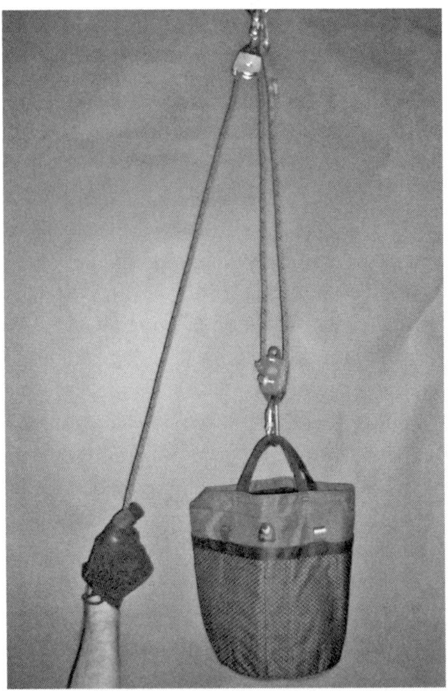

Figure 7.33 A 3:1 mechanical advantage (Z-rig).

Figure 7.34 Commercially available pre-rigged haul system. Credit: Pigeon Mountain Industries, Inc.

All of the above-described systems are "simple mechanical advantage." Stacking multiple systems on top of one another or integrating multiple systems into one another creates compound or complex mechanical advantage systems. Such systems are sometimes used by professional rescuers, but are beyond the scope of this book.

While haul systems may be created as needed during a rescue operation by using pulleys, connectors, and rope grabs, the most straightforward approach for workmate retrieval is to simply purchase pre-rigged haul systems from reputable life safety equipment manufacturers. This helps to ensure compatible components, and prevents workers from having to think through the complexities of mechanical advantage during a life-saving moment.

A pre-rigged haul system generally consists of a lightweight rope, two multi-sheave pulleys, and a *progress capture device (PCD)*. Commercial, pre-rigged systems are usually set up to provide either a 4:1 mechanical advantage or a 5:1 mechanical advantage, depending on which pulley you place at the "top." These may come packaged in an easy-to-carry bag (Fig. 7.34).

Creating mechanical advantage is just one aspect of a raising system. As you pull the load upward, the pulleys will move toward one another until eventually they are bunched together (Fig. 7.35). Unless the subject must be raised only a very short distance, several resets will most likely be required. To accommodate resets, you will need to devise some method of capturing the progress gained during each haul evolution.

One method to achieve this in a piggy-backed system is to place a rope grab on the main line at the anchor and reset it with each pull. This is known as a PCD, or simply as a "ratchet." This progress capture rope grab can also serve double duty as a backup safety mechanism in the event that the person(s) hauling should let go.

Figure 7.35 Pulleys move toward one another as the system is pulled.

More specific examples and guidance for implementing PCDs are provided in Chapter 9.

RESCUE SYSTEM BELAYS

Whether raising or lowering, consider whether to incorporate a belay, or backup, into your rescue system. The term belay means, simply, to stop or arrest. In Chapter 6 we discussed personal belay systems for ascending, descending, and climbing. A rescue belay is similar, but usually for a heavier load. It generally consists of a secondary system used as a backup to catch the load in case of a failure in the primary system.

Not all rescue systems require a belay, and in some cases a belay could put rescuers at additional hazard (e.g., where primary and secondary systems are likely to become entangled or create a snag, or in moving water where a belay could easily be inadvertently activated and leave the rescuer[s] and/or subject[s] in a precarious position).

Careful consideration should be given to whether a given situation warrants the additional time, equipment, and personnel required for a belay. A belay should definitely be considered where:

- There is a high likelihood of failure of the main system.
- The primary raising/lowering system being used for rescue does not automatically stop/lock if the user lets go.
- The protection provided by the belay outweighs the potential hazards that the belay might create.

Belays commonly used in rescue include *active belay* and *passive belay* systems. An active belay usually consists of a secondary brake device operated by a belayer. A passive belay is one that grabs automatically with minimal human intervention, such as a rope grab.

In an active belay system, the rescuer is tied directly in to a rope, the running end of which is passed through a braking device that is anchored to a solid point and managed by a second rescuer. This second rescuer would be known in this case as the "belayer." As the first rescuer moves, the belayer maintains light tension on the "belay" rope. If at any time the primary system of the first rescuer fails, the belayer would increase tension on the belay rope to prevent it moving through the brake, thereby catching the fall.

A major disadvantage of this system is that it is completely reliant on human performance. Especially in co-worker assisted rescue situations, where rescue is neither the vocation nor the primary responsibility of responders, complete reliance on human reactions is not the best choice.

At the other extreme is the passive belay. The fall arrest block, also known as a self-retracting lifeline (SRL), is shown in Figure 7.36 as one example of a passive belay. An SRL generally consists of an enclosed case with a cable or fiber line is wound around a tensioned reel (or "block") inside the case. The block is fitted with a centripetal brake that will lock off in the event of sudden impact. Some SRLs feature the block fixed directly to the dorsal D-ring, with the line anchored above, while others are designed for the block to be attached to an anchorage above the user. As the user moves, the tensioned reel pays out and retracts, allowing the cable to travel as long as there is no sudden increase in velocity. Should the person fall, the SRL will automatically engage, stopping the fall. For this reason, care must be taken when using an SRL to not move too quickly, or the device will lock off. Once locked, tension must be removed from the device in order to release the brake.

While an active belay is arguably less secure than a passive belay, it does offer more flexibility in use, particularly in complicated environments, and permits a modicum of control by the belayer, enabling him or her to provide more assistance during a rescue—up to and including even lowering.

A good compromise is a hybrid system wherein the device is operated by the belayer but engages automatically when the rope or line is pulled rapidly, such as in a fall (Fig. 7.37). With such a device, the belayer can quickly feed rope in either direction, maintaining versatility, but if the working end of the rope is pulled rapidly

Figure 7.36 Self-retracting lifeline. Credit: Reliance Industries, LLC.

Figure 7.37 Manually operated, automatically locking belay device.

a secure locking mechanism will engage. Manually operated, automatically locking belay devices generally feature mechanisms by which, if necessary, the belayer can release the tension of the locked off device without having to lift the load.

Professional rescuers sometimes use tandem prusik belay systems. This type of system has significant limitations, and can be unpredictable in performance based upon the relationship between the rope and cord, wet/dry conditions, and other factors. For this reason, tandem prusik belay systems are seldom considered appropriate for co-worker assisted rescue in the workplace, and are beyond the scope of this book.

SAFETY FACTORS

Having reviewed personal equipment in Chapter 5, ream rescue equipment in Chapter 6, and principles of rigging in this chapter, it is important to emphasize here the role that safety factors[3] should play in system design and use.

Clearly the strength of equipment must be at least somewhat greater than the amount of force that we expect to exert on it. Recall that we discussed this concept previously in terms of "minimum breaking strengths" and "design loads." The same holds true of that equipment as it is rigged into a system; the system should be built to withstand greater potential force than the actual force expected on it.

Many people think of a safety factor as being the difference between the strength of a component (e.g., a rope) as compared with the load that is expected to be placed on that component. While this number, expressed as a ratio, is an important number, by itself it is known only as a "load ratio" (Table 7.2) and by itself is not a particularly useful piece of information.

TABLE 7.2 Load Ratio Example

Component Load Ratio
Rope Strength : Rescuer + Subject Weight
6,000 lbf:600 lbs
10:1

Knowing the load ratios of each component in your system is nice, but only by understanding the actual strength of each component as it is interconnected with other components is it possible to determine the true safety factor of your system.

As we have discussed throughout this book thus far, there are many things that can serve to amplify forces imparted by a load, including friction, angles, and the mechanics of the systems themselves. Consider all these things in realizing that

[3]In common rescue terminology, safety factors are usually expressed as a ratio. While the term "factor" technically mathematically represents a *multiplier* rather than a *ratio*, the term *safety factor* is acceptable common usage. We can accurately use the term "safety factor" because the ratio is always relation to 1. Saying 6:1 merely represents a factor of 6 as applied to the weakest point in the system.

the difference between maximum anticipated load at any given point in the system, as compared with the weakest point in the system, is what is really important.

A system is only as strong as its weakest link, but the weakest link in a system may not be—in fact probably isn't—an individual component at all. Rated equipment strengths can be affected by rigging methods, quality of anchors, and force multipliers both within and on the system. Often the weakest point in a system will be the interface between two components, or between a component and an external force (such as an edge).

The ratio between the breaking strength of the weakest point of a system as compared with the maximum anticipated load constitutes what we know as a safety factor.

A System Safety Factor May Be Calculated Using the Following Formula:

$$S / L$$

where

S = strength of the weakest point in the system, and

L = maximum anticipated load.

To use an example from the fall protection world, many regulatory requirements specify that a fall protection system should be capable of withstanding 5,000 lbs of force. Standards and regulations also specify that an arrested fall should mitigate the force to no greater than 1,800 pounds during a fall arrest. By dividing 5,000 by 1,800, we learn that such a system has a 2.7:1 safety factor. That is, the system is 2.7 times stronger than the maximum force that the load is reasonably expected to exert.

Every system should be built to withstand greater potential force than the actual force expected on the system. How much stronger will depend on the situation. A situation where failure is unlikely and the consequence of a failure, even if it did occur, would be minimal, may justify a very low safety factor. In contrast, in a system that is subject to known conditions that may cause it to fail and/or where a system failure is likely to be catastrophic to the people using it, a higher safety factor may be in order.

Safety factors will vary depending on circumstances, and the target safety factor for a given operation should be established by a Qualified or Competent Person. This is one case where a higher number is not always necessarily better. The goal is to achieve as high a system safety factor as is needed, and that is reasonable to rig in the given conditions with time restraints, equipment constraints, and unique location circumstances that are often beyond our control. Mandating an extremely high safety factor could backfire by making equipment unnecessarily heavy or unwieldy, thereby potentially increasing the probability of failure.

With the target safety factor in mind, a Competent Person should carefully assess the actual safety factor once the system is built, to see how it compares. If necessary, the safety factor of a given system can be easily altered either by adjusting the system itself to use stronger components or improve the relationships between components, or by resolving to minimize loading potential on the system (e.g., one person instead of two, shorter fall potentials, etc.)

A system safety factor generally refers some multiple of the anticipated load, intended to give some margin of safety in the event of unforeseen events. Establish your desired system safety factor first, then determine whether your system measures up. Safety factors are not an exact science, and should be determined by knowledgeable, experienced, and skilled personnel. Precision and accuracy will depend on several things, including how accurately you estimated the maximum load, how accurately you pinpointed the weakest link, and what the consequences might be in the event of a failure, among other things.

SUMMARY

There are many factors to consider in rigging systems for co-worker assisted rescue. By understanding the basic principles of gravity, friction, and angles, the responder can increase his or her ability to cope with unexpected circumstances and utilize equipment safely and effectively.

POST-CHAPTER QUIZ

Chapter 7

1. The resistance of an object to the medium through which or on which it is traveling is known as
   - a) Gravity
   - b) Force
   - c) Friction
   - d) Energy
2. Friction is imparted on a rope in a rescue system by
   - a) Surfaces over which the rope travels
   - b) Descenders and other hardware
   - c) The grip of the rescuer's hand
   - d) All of the above
3. The best choice for protecting a moving rope over an edge is by using
   - a) An edge roller
   - b) A canvas edge pad
   - c) A plastic edge protector
   - d) Another rope

4. The strength of an anchor used for co-worker assisted rescue should be at least 3,000 lbs multiplied by
   - a) The force of gravity
   - b) The number of people who will be attached to it
   - c) The strength of a fall protection anchor
   - d) 5,000 lbs
5. Increasing angle between legs of a two-point load-sharing anchor system increases the force on each one so that at 120° each anchor sees 100% of the load. With a 175° angle between legs, each anchor can see up to
   - a) Half the load
   - b) 100% of the load
   - c) Twice the load
   - d) Eleven times the load
6. As the internal angle of a bend increases, as in a deviation anchor, the forces on the anchor or other equipment at that bending point will
   - a) Increase to up to ten times the actual load
   - b) Decrease to half the actual load
   - c) Increase to up to two times the actual load
   - d) Decrease to one fourth of the actual load
7. Whenever the first pulley in a mechanical advantage system is placed at the load, the result is an
   - a) Even numbered mechanical advantage
   - b) Odd numbered mechanical advantage
8. A 3:1 mechanical advantage system is also sometimes referred to as a
   - a) DeRig
   - b) Compound system
   - c) Pulley
   - d) Z-rig
9. A belay system wherein the device is operated by a belayer but offers an automatically locking function when the rope or line is pulled rapidly is known as
   - a) An SRL
   - b) A manually controlled auto-locking system
   - c) A free running system
   - d) Not a belay at all
10. The difference between the strength of a component (e.g., a rope) as compared with the load that is expected to be placed on that component is known as
    - a) A fall factor
    - b) A system safety factor
    - c) A mechanical advantage ratio
    - d) A component load ratio

11. The difference between the weakest point in a system as compared with the maximum anticipated load at that point, expressed as a ratio, is known as

a) The system safety factor

b) Minimum breaking strength

c) A Z-rig

d) Mechanical advantage

ANSWER KEY

1. c

2. d

3. a

4. b

5. d

6. c

7. a

8. d

9. b

10. d

11. a

CHAPTER 8

SINGLE RESCUER TECHNIQUES

When you have completed this chapter you should understand several methods and techniques used for rescue by a single rescuer including:

- Anchors for rescue.
- Reaching the subject from above.
- Reaching the subject from below.
- How to transfer a subject to a rescue system.
- How to release a subject from a suspended system.
- How to raise and/or lower a subject to safety.

Rescue of another person is a responsibility that requires advance preparation of tools and methods, as well as clear thinking during the incident. This chapter builds on the personal skills addressed in Chapter 6 and the rigging skills covered in Chapter 7 to provide practical guidance in techniques and methods that an individual responder can use during an emergency. The methods described in this chapter can be performed by one person, while the methods described in Chapter 9 may be more appropriate when several responders are available.

As with other material in this book, the information provided here is not intended to replace hands-on skills training. While this text is a useful resource and provides a good foundation for understanding, there is no substitute for tutelage by a competent instructor. This text will address specific rescue methods applicable to commonly encountered types of fall protection. Similar concepts may also be applied by trained and experienced persons to situations that may not be specifically addressed here.

When a co-worker is called into action to assist a workmate who has fallen, chances are the workers know one another on a personal level. They may even be friends. The psychological stress this can add to the situation cannot be overemphasized. One thing we do *not* want to do is make things worse. To mitigate the negative impact that stress and adrenaline could have on the responding worker, individuals responsible for co-worker rescue must be very familiar with the methods and techniques they will use in assisting the subject. The tasks that the assisting worker will be using probably vary significantly from his or her daily work. The familiarity and skill required to overcome an adrenaline-pumped mind requires training and practice.

Falls from Height: A Guide to Rescue Planning, First Edition. Loui McCurley.

WHEN A WORKER FALLS

A fallen subject generally fits into one of two categories:

1. Either he or she was not properly using appropriate fall protection equipment and has hit the ground or other obstruction, or
2. The subject was caught by some means of fall arrest, and remains in this "caught" position.

In the case of the first category, if the subject has hit the ground, he or she may or may not be injured. Statistically, falls from a height to ground tend not to have great outcomes, often ending in a fatality. The extent of the subject's injuries will depend on the distance he or she fell, the surface upon which he or she landed, time lapse between incident and rescue, and other factors. Even if the subject does not appear to be, or does not think he or she is, badly injured, medical attention should be rendered to verify the absence of life threatening internal injuries.

While intensive medical intervention is usually not within the scope of co-worker assisted rescue responders, basic lifesaving skills such as CPR and first aid can help a responder to recognize and correct some immediate, life threatening injuries. Given the economical availability of basic first aid and CPR training from organizations such as the Red Cross, American Heart Association, and National Safety Council, at least this level of instruction is highly recommended for any workers who might be likely to find themselves in a co-worker assisted rescue role.

When a subject hits the ground from a significant distance, blunt trauma is induced and life threatening injuries are quite likely. Stories such as that of the steel worker whose misstep on a seventh story installation sent him crashing through an atrium and onto the floor below amid stunned hotel guests[1] abound. In such situations, immediate medical response is required, though unlikely to result in a positive outcome.

A fall may be more survivable if the subject falls not to the ground, but onto obstructions below. Such was the case when a 31-year-old Austin, Texas, tower technician fell 100 feet onto a waveguide bridge below, and lived.[2] Other examples may include falling to another level, into a safety net, onto machinery, into a tank, or even onto support pieces of a structure. The employer's emergency action plan should consider these possibilities, and should specify what to do in each case. A subject who has fallen to a different level poses additional hazards to co-workers and others who may be motivated to rush in to assist whether or not they are properly prepared. Ensuring that there is a plan and that co-workers are prepared to initiate it will help to avoid further injury by spontaneous responders, and will help bring the incident to positive conclusion.

In the event that the subject's fall is stopped by some preplanned method of fall arrest, such as a lifeline system or a lanyard, it is quite likely that he or she is

[1]"Worker Falls to His Death at Courthouse Hotel," ARLnow.com, July 21, 2010, accessed July 24, 2012.

[2]"Climber's Fall on AT&T Project May Have Forced Goodman Networks' Unconventional Stand-Down," Wireless Estimator, http://www.wirelessestimator.com/breaking_news.cfm, May 25, 2012 (accessed July 24, 2012).

uninjured but simply suspended out of reach. The complexities of accessing and extricating such a person can be reduced with a little forethought and practice. When a subject's fall is properly arrested, even if he or she may have received some injury during the fall, co-worker assisted rescue is most likely to result in a good outcome.

The first preferred method of response is always for the subject to escape from the predicament himself. Techniques for this are addressed in Chapter 6. When personal escape methods are not successful, the response should escalate to the next level: co-worker assisted rescue. This chapter focuses first on the height-oriented aspects of how one co-worker might remove a subject from a reasonably accessible different level, and then later in the chapter we will discuss methods that a lone co-worker might use for retrieving a subject who is suspended in a fall protection harness.

The matter of extrication and rescue are sometimes compounded by other hazards, such as machinery, entanglement, confined spaces, collapse, and other special environments and conditions. Because these possibilities are so diverse, it is not possible to thoroughly and adequately address all combinations of possibilities in this text. These (and other) hazards should be identified as part of the rescue preplan discussed in Chapter 2. A brief introduction to hazards found in a few common workplace environments is provided in Chapter 10, but this is merely an overview. Anyone responding to an at-height emergency that involves additional hazards should be specially trained in those areas as well as in height rescue. A responder who is not fully qualified in all hazards associated with a situation in which he finds himself should not respond, but should instead call upon, and wait for, additional assistance.

Situations involving several known hazards existing simultaneously while work is being performed often provide the best opportunity for utilization of a professional standby rescue team whose skills are primed and honed for the combination of hazards found in that place.

RESPONDING TO A FALL

Local emergency services should be notified and activated as soon as possible following any potentially life threatening incident, including a fall, even if the fall is successfully arrested by some manner of fall arrest. The emergency rescue plan outlined in Chapter 2 should include details regarding who to contact in the event of an emergency, including notification of security, management, internal responders, and external emergency services. The plan may even offer specific guidance for who to call for which kind of incident, applicable phone numbers, approximate anticipated response times, and an alternative plan in case the first preferred action is not feasible.

Part of effective response training includes practicing this notification process. Personnel should specifically request both rescue and medical response when contacting external emergency services, and should stay on the phone with the emergency call center to provide key information. Information that they should be prepared to offer includes the caller's name, a return phone number, and an address/location of the incident. Be sure to include specific instructions for responders—which door/gate should responders enter, whether someone will meet them, whether security has been notified, and so on. The dispatcher will also want as complete a description of the incident as possible, and the caller should request both rescue and medical response. The request for rescue should be as descriptive as possible: How long ago did the incident occur? Is the subject still suspended in the air? How? How far off the ground? Did he or she hit any obstacles? Are there other hazards (chemicals, confined space, etc.) involved? Do you know if there are any injuries? How old is the individual? Is he or she conscious? Has verbal contact been established? If you are not prepared to answer all of these questions right away, don't worry; call anyway, and just provide as much information as possible. The sooner you call, the better.

Even if it is assumed that personal escape or assisted rescue will be successful, professional rescue should still be initiated immediately. If co-worker assisted rescuers have been able to safely retrieve the subject by the time rescue arrives, so much the better: the rescuers can go home, and everyone will be happy. On the other hand, if neither personal escape nor assisted rescue has been successful by the time professional rescue arrives, the precious minutes saved by an early request for assistance could save the subject's life.

Any further action taken by co-workers should be within the scope and only to the extent of their prior training and approval from their employer. A good post-fall rescue plan will recognize the existence of other workers at the site, and will provide specific guidance as to what actions those workers should take during the response phase of an incident response. Even workers who are not directly responding or assisting with the incident are likely to be shaken up by the ordeal and should not be expected to simply go on with "business as usual." For additional review on this subject, see Chapter 2.

As a co-worker assisted rescue responder, your first priority should be your own safety. There is no profit in becoming a liability to the situation yourself, so follow the rescuer's safety maxim:

- My own safety first
- The safety of my co-responders second
- The person we are here to rescue last.

In keeping with this, you should take a few moments to perform a quick site evaluation as soon as you arrive at the scene of an incident. This is not the time to begin preparing a written document, but to consider the situation at hand and how your written preplan might apply to it. You should be prepared to communicate key points that you observe about the incident to other responders and/or the prevailing incident manager. Important details include anything related to scene safety, location/condition of the subject, limitations at the site (hazards, visibility, communications, access, etc.), and other factors that may affect the development of the operation.

Steps for Co-Worker Rescue:

1. Call emergency services.
2. Ensure that the scene is safe.
3. Make verbal contact with the subject as soon as possible.
4. Prepare to assist.
5. Initiate assistance.

As soon as the scene is deemed safe, you can turn your attention to reaching and retrieving the subject. Your first goal should be to make contact with the subject, at least verbal contact, immediately after the fall. During the moments after a fall the subject may be shaken up and not thinking clearly, and he or she may or may not be injured. If the fall was caused by a medical event, or if he or she hit an obstacle or was impacted by some other hazard, he or she may even be unconscious.

If the subject is conscious, help him or her to calm down. Ask whether he or she is injured, and whether he or she is able to help him- or herself. Help the subject to determine whether he or she is physically and psychologically able to extricate him- or herself using the knowledge, skills, and equipment he or she already has. If the subject is unconscious, badly injured, extremely emotionally distraught, or otherwise incapable of personal escape, initiate assisted rescue procedures immediately. If the subject is not injured and if he or she is adequately trained, prepared, and capable of personal escape, he or she should be encouraged to immediately begin to implement methods such as those described in Chapter 6, to the extent that he or she is able.

Sometimes, even a well-trained subject who has successfully performed personal escape skills during practice may be incapable of personal escape during an emergency simply due to distress. During the personal escape attempt, an experienced person should act as an observer, with the assigned task of watching every move that the subject makes. If the subject begins to do something dangerous or inappropriate, it may be that he or she is not able to perform the necessary skills at that moment, and the observer should encourage him or her to stop and wait for assisted rescue.

This is not an appropriate time for the observer (or anyone else) to provide coaching, advice, or rescue instruction to the subject. The subject either is able to extricate him- or herself using the knowledge, skills, and equipment available at that moment, or he or she is not. Shouting instructions from afar will only frustrate everyone, will add stress to the situation, and is not likely to result in a good outcome. If the subject cannot remember how, or is unable to escape on his own, he or she should make him- or herself as comfortable as possible and wait for assistance.

Even while the subject is attempting to implement his means of personal escape, other co-workers should activate local emergency services, and should begin to implement the next level of response: co-worker assisted rescue. Whether this is a ladder, a bucket truck, a rope-based system, or some other means, it should be employed soon as possible. This might involve bringing a rescue kit or other equipment from its stored location, and perhaps even beginning to set up an anchorage(s) and/or system(s) for the assisted rescue. Setup should take only a few short minutes. By this time the subject will either have succeeded in personal escape, or will be worn out from trying.

If the subject has not self-extricated by the time assisted rescue is ready to be put into action, the responsible person onsite will have to make a judgment call as to whether it is best to allow the personal escape attempt to continue, or for the subject to stop what he is doing and wait for assistance. This determination will be based at least in part on the resources available, the condition of the subject, and the expected time/likelihood for achieving success.

Reach the Subject

The best method for co-worker assisted rescue is one that co-workers can use to safely and effectively reach the subject without creating additional hazard or harm. It must be readily available, it must be appropriate to the situation at hand, and co-workers must be trained to apply it. If any of these is not the case, it is best to wait for professional rescue.

There are three key steps involved in retrieving a fallen subject:

1. *Reach* and *secure* the subject
2. *Release* the subject from his or her predicament, and
3. *Extricate* the subject to a secure location.

Always plan first to use the simplest approach available to provide assisted rescue. Even if co-workers have been trained extensively in technical rope-rescue methods, consider whether a nearby ladder or bucket truck might be available, and whether such a method may be the best choice.

Platforms and Working Surfaces Accidents can happen even on platforms and working surfaces where workers are protected by guardrails or restraint systems. Unfortunately, in these cases a fall often results in the subject impacting the ground or a lower surface. The disadvantage of hitting the ground is obvious; impact can be catastrophic. If there is an advantage, it may be that the subject is more likely to be readily accessible via stairs, walkways, or other walking/working surfaces, thereby enabling responders to reach him or her more easily.

Not all surfaces on which a subject might land are readily accessible, however. Even so, with a little preplanning, rescue of a subject who has fallen to a lower level can be readily achieved in a few simple steps. The first step, as always, is to reach the subject. If the subject is attached to fall protection, his or her lanyard or lifeline may provide the means of reaching him or her—for example, by using the vertical lifeline to which the subject is attached as the means of rescue. When a rescue system is built directly in to the system to which a subject is already connected, this is called an *in-line* rescue system.

A Suspended Subject A subject who has fallen into a safety system is very likely locked into a suspended condition by the force of gravity working in opposition to the safety mechanism that arrested his or her fall. This may be an activated fall arrest rope grab, a centrifugal braking mechanism in a self-retracting lifeline, or simply an anchored lanyard. Releasing the subject from this locked-on condition can be a difficult proposition.

While cutting the lanyard with a knife or scissors may seem like a logical solution, this is not the first most recommended approach. Some lanyards cannot easily be cut, and in any case cutting a subject loose opens a Pandora's box of new hazards that may create a worse situation than the one that is resolved. Clearly the most obvious concern is that the slightest gaffe could result in accidentally cutting the rescue line instead of, or in addition to, the fall arrest line. This could result in catastrophic failure.

Aside from this, consider also the continued effects of gravity against the suspended load. Even if the rescue/retrieval line is pulled reasonably taut, simple physics dictates that cutting the line that is under tension will result in a sudden loading of the line that is less taut. Even if the rescuer has been able to pull the rescue line quite tight, a drop of several inches and a potential force of near twice the load is quite likely. For a person who has already endured an unpleasant fall, this is at best an unnecessary shock, and at worse could precipitate a potentially injurious secondary event.

Raising the subject, even just a few inches, will facilitate the release of the locking mechanism, avoiding the hazards of cutting altogether, and is a much better option if available.

Non-Entry Methods The preferred method of co-worker assisted rescue will usually be one that does not require the rescuer to enter the hazard him- or herself, but instead allows the rescuer to perform the rescue from a safe zone. Reaching a fallen subject without entering the hazard can be simplified by preplanning rescue into the fall protection system. Whenever possible, incorporating post-fall retrieval capabilities into a system to which the subject is already attached in advance can be very advantageous in that all three aspects of retrieval (reach, release, and extricate) can usually be combined and simplified.

Some work environments are conducive to pre-rigging non-entry rescue systems so that a rescue can be effected with little or no additional equipment or rigging time. If the system is designed and approved to do so, it may double as restraint, or even limited fall arrest, while at the same time providing a backup capability.

One example of this would be a self-retracting lifeline (SRL) system (sometimes referred to as a fall arrest block) that incorporates an integral rescue mechanism. Some such mechanisms deploy automatically, serving as a controlled rate descender (CRD) over which the user has no direct control. Others incorporate a winch mechanism that allows a co-worker assisted rescue responder to raise and/or lower the subject in an emergency.

SRL/CRD/retrieval systems are most useful on frequently accessed sites where the block can be anchored nearly or directly above the work location. In the event of an incident, rescue is as simple as turning a crank to raise the person to safety.

A similar effect can be achieved through the manual belay methods discussed in Chapter 7. In this type of system, the worker attaches to a safety line, such as a rope, which is in turn reeved through a belay device that is attached to an anchorage. A trained co-worker can then manage the system to provide extension and retraction of the safety line as the worker moves through the work area. In the event that the worker becomes incapacitated, the co-worker can use the belay line to lower or raise him or her to a secure location using techniques as outlined in Chapter 7.

If the worker will be using a lifeline, an integrated rescue capability can be built directly into the lifeline in advance by using twice as much rope as necessary, and anchoring it with a tensionless anchor as shown in Chapter 7.

With enough spare rope in the system behind the anchor, if the worker should fall anywhere on the line, he or she may be lowered to ground simply by releasing the anchor and using the tensionless anchor as friction to lower the load under control. A similar effect may be achieved by integrating a braking device into the anchorage and leaving enough standing rope behind the anchor to effect a complete lower.

This is an excellent choice where suspended rope work is being performed, but one significant limitation of this type of workmate retrieval method is that the amount of rope required is at least double what would be required for the drop itself. In the event that the subject falls near the top of the drop, enough spare rope must be available at the anchor to lower the subject all the way to the ground. Therefore, this method may be quite impractical where significant heights are involved.

Another option is to attach a raising system to the lifeline. This can be as simple as attaching a pre-rigged haul system to a sturdy anchor at the top, and connecting it to the lifeline with a substantial rope grab. This is known as a "piggyback" system.

Pulling on the piggybacked haul system raises the lifeline, and the subject (if attached) follows. This technique may be used to lift the subject off a lower level, and even to haul the subject all the way up to safety, or simply to release the subject from a stuck condition. Keeping the haul system in line with the direction that you want the load to move will make it more efficient. Having a haul system cocked off at an angle results in wasted energy.

Depending how far the subject needs to be hauled, it may be necessary to collect the slack in the lifeline and lock it off to capture progress so that the haul system may be reset, as described in Chapter 7.

If the subject is not pre-attached to a vertical lifeline, but falls directly onto a lanyard, force absorber, or SRL that is itself attached to an anchorage, the haul system may be attached directly to that fall protection system in a similar manner. Making the connection to a lanyard or SRL can be a challenge, however, because many rope grabs will not effectively grip the cable or webbing of which fall protection lanyards are made. Be sure to consider this in the rescue preplan.

Allowable fall distances vary depending on regulatory requirements, industry sector, and type of equipment used. In addition to actual fall distance, most force absorbing systems allow a certain amount of "extension" in absorbing force including up to 3.5 feet of allowable deceleration distance and the natural settling-in, or stretch, in the system. This means, for example, that a subject who might have fallen a permissible 6 feet from a normally quite accessible platform may actually be the equivalent of one to two stories below the platform, and well out of reach, after the fall.

Depending how far away from rescuers the subject is, it may be possible to connect a rescue line directly to his or her harness. If the subject is ambulatory, allowing him or her to assist with this is a perfectly acceptable solution. Simply lower the rope end (with a connector) to the subject, and ask him or her to clip it to an appropriate rescue point on his or her harness, such as a front D-ring.

If the subject is not ambulatory, a specially designed pole (Fig. 8.1) may be used to extend the rescuer's reach and attach a connector directly to the harness of the fallen subject, or to an appropriate connector in the fall arrest system. This method, sometimes called *remote connection*, requires some practice and a steady hand, especially in consideration of the fact that some such poles extend to lengths upward of 10 feet (3 meters). These poles must also be fitted with a mechanism for attaching the connector at the far end of the pole and holding it open until it is connected to the subject.

A reach pole is used as follows:

1. Connect the pole with a safety lanyard to the rescuer or an anchor, to prevent accidentally dropping it.
2. Attach the connector (with rescue line attached) to the end of the telescopic pole, with the gate held in the open position (Fig. 8.2).

Figure 8.1 A reach pole. Credit: Vertical Rescue Solutions/PMI.

Figure 8.2 Use a special clip to hold the connector open. Credit: Vertical Rescue Solutions/PMI.

Figure 8.3 Extend pole segments one at a time. Credit: Vertical Rescue Solutions/PMI.

3. Starting with the end of the pole that will eventually be closest to the subject, extend each segment of the pole toward the subject, locking the segments into place in turn as they are extended (Fig. 8.3).
4. With a steady hand, hook the connector to a suitable D-ring or other appropriate load-bearing component on the subject's harness (Fig. 8.4). Using the same attachment point that is already under load is usually the easiest, but first verify that this attachment point has not been compromised during the fall.
5. When you are certain that the hook is in place, release the pole from the connector (Fig. 8.5). Normally, this is accomplished with a sharp pull upward; the connector gate should automatically close upon release.
6. Verify that the life safety rope is securely in place, and pull all slack out of the system.

Figure 8.4 Clip the connector to the subject's harness. Credit: Vertical Rescue Solutions/PMI.

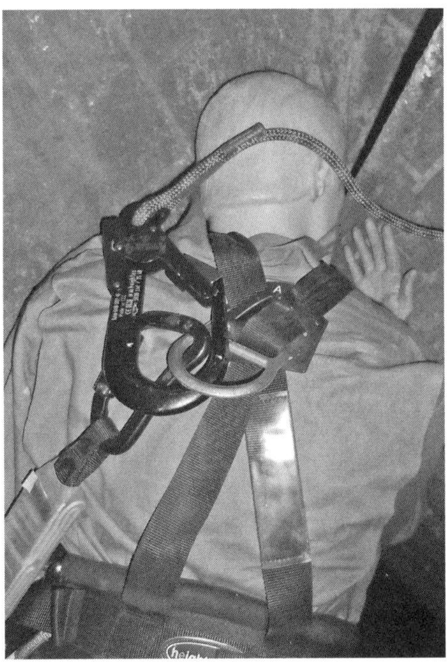

Figure 8.5 Release the pole from the connector. Credit: Vertical Rescue Solutions/PMI.

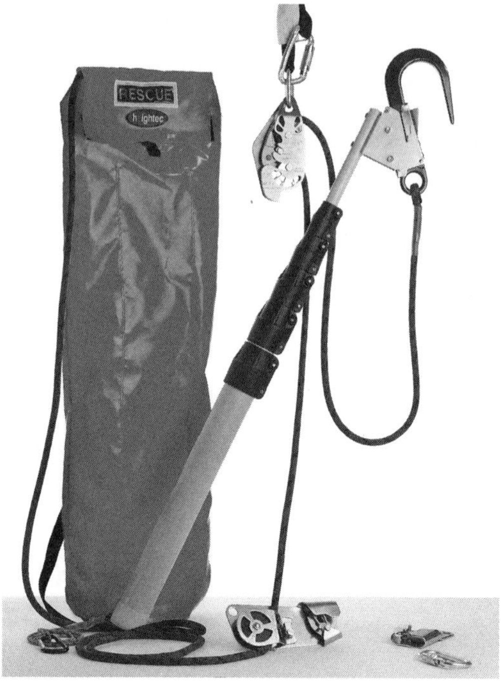

Figure 8.6 All-in-one co-worker rescue solution with raising and lowering capability. Credit: Heightec.

Rescue Solutions Although professional rescuers often build rescue lowering and raising systems from scratch, using a variety of ropes, pulleys, connectors, or other components from multiple sources, this approach is generally discouraged for use by co-worker assisted rescuers. It is assumed that anyone assigned to co-worker assisted rescue is not likely to be employed as a professional rescue technician, and as such it is not realistic to expect this person to be able to assemble rescue systems from memory, using a hodgepodge of gear.

If you are assigned or volunteer as a co-worker assisted rescue responder, it is wise to request of your employer a manufactured rescue kit from a reputable manufacturer, to secure training in the use of that kit, to practice regularly, and to re-train on that kit periodically. For a complete co-worker rescue solution, consider using an all-in-one system that facilitates lowering and then easily converts the main rescue line into a haul system as needed, such as that shown in Figure 8.6.

Equipment in kits and solutions should be compatible, designed for interoperability, and tested together as a system for greatest security. Those with few removable components, and that convert easily from lower to raise and back again to lower, offer the greatest versatility and least potential for misuse.

To use an in-line haul system to raise the subject enough to release them from fall arrest:

1. Select an anchor with at least 3–6 feet of "working space" between it and the edge. Connect an anchor, a pulley, and a progress capture device (or use a multi-function device to act as both pulley and progress capture).

Figure 8.7 Reeve the rope through the pulley to create a 3:1 mechanical advantage. Credit: Vertical Rescue Solutions/PMI.

2. Connect a rescue rope to the subject's fall arrest system using a rope grab or reach pole.
3. Attach a pulley to the rope between the anchor and the edge, but as near to the edge as possible without having it drag on the ground or connect with other obstacles.
4. Extend the tail of the rope from the anchored pulley toward the pulley that is near the edge, reeve it through the pulley, and then back toward the anchor. This creates an in-line 3:1 mechanical advantage (Fig. 8.7)
5. To begin the uphaul, pull steadily and carefully on the leg of rope that comes off the "outside" of the pulley, taking care to keep a brake hand on the rope as you pull. Your progress capture device should permit the rope to run through the system in an upward direction, but prevent loss of progress when the rope is released.

The subject should need only to be pulled a short distance in order to release his or her fall protection mechanism. If the lifeline system anchorage is accessible, simply unclip it as soon as there is sufficient slack to do so. An SRL brake will usually automatically disengage when tension is released, rendering disconnection unnecessary.

Once the original fall arrest system has been released, the subject may be extricated to a secure location.

Extricate the Subject to a Secure Location

The final phase of workmate retrieval involves extrication of the subject to a secure location. Because a fallen worker is generally either suspended or resting on an intermediary point, extrication usually involves either lowering or raising the subject

Figure 8.8 Lifting a subject over an edge can be quite difficult. Credit: Vertical Rescue Solutions/PMI.

to the ground level or a platform. If he or she will be raised all the way to the top, it may not be necessary to disconnect the fall arrest system unless entanglement potential exists.

As was noted in our discussions of releasing the subject from suspension, some means of raising the subject may be required to release him or her from fall arrest. With a raising system already in place, it may be appropriate to continue to raise the subject all the way to the point of anchorage. In fact, where obstructions exist below (such as open water, machinery, or other hazards), raising may be the *only* option.

On reaching the top, the casualty may need to be manhandled up and over the edge (Fig. 8.8). This can be the most difficult part of raising a subject. An undercut or sharply angled edge will make the edge transition even more difficult. Rigging a high directional, as discussed in Chapter 7, can help to facilitate edge negotiation.

Sometimes it is more sensible to lower the subject. When it is necessary to raise and then lower a subject during the same operation, it is best to use a multi-function device, as described in Chapter 5, as the pulley/progress capture in the haul system. Placing a multifunction device in this position allows quick conversion from raise to lower, and reduces the potential for error in this exercise.

To convert a raise to a lower using a multi-function device:

1. Follow the procedure to release the subject from their fall arrest system and raise them onto the rescue system.
2. While maintaining a firm grasp on the tail of the rope where it exits the braking device, remove the haul system from the rescue line.
3. Make certain there is enough rope to lower the subject as far as needed to effect the rescue, and that the descent path is clear and free of any obstructions.

4. Activate the descender (per device instructions) to lower the subject in a slow, steady, and controlled manner.

This method is *only* appropriate where an actual lowering device is part of the system. Haul systems that incorporate only pulleys and camming devices, and not a braking device, are intended primarily for raising, and should never be used as the primary friction mechanism to lower a load any great distance.

METHODS FOR WORKMATE RETRIEVAL: ENTRY METHODS

The methods described in the preceding pages work for many situations, but what do you do in cases where the subject cannot easily be reached by one of these methods? There are times when the best choice for retrieving a fallen workmate requires the co-worker assisted rescuer to enter the vertical environment to reach the subject. Sometimes this is because it is the *only* possible choice, while at other times circumstances simply make such methods the *best* choice. This approach may be appropriate when the subject needs immediate medical attention, where extrication is difficult, if the subject is not otherwise accessible, if the rescue system cannot be connected remotely, or if the subject cannot otherwise be readily disconnected from his fall protection.

This is a decision that requires good judgment on the part of the rescuer. Entering the vertical environment is not something to be undertaken lightly and this type of co-worker rescue should only be performed by those who are specifically trained in appropriate methods, where a preplan has been established and prepared for this approach, and where it is the most appropriate alternative.

Forethought and adequate preparation are necessary to offset the potential disadvantages of the rescuer entering the vertical environment in order to effect the retrieval of a workmate. The primary disadvantage is of course the verticality itself. Any environment that has provoked one fall certainly has potential to invite another. Would-be rescuers should never expose themselves to a potential fall hazard without adequate protection. This dictum holds true in relation not just to the fall hazard, but also to other hazards including environmental, chemical, and respiratory.

When retrieving a workmate using entry methods, the three key objectives remain to:

1. Reach and secure the subject,
2. Release the subject from the predicament, and
3. Extricate the subject to a secure location.

A rescuer may opt to enter the vertical environment strictly for the purpose of reaching the subject, and then may choose to set an anchor and lower the subject from that point, or the rescuer may choose to secure the subject to his or her own system and continue with the subject to safety. This latter approach is known as *pickoff rescue*.

Figure 8.9 Carry the rope in a bag when descending for rescue.

Reach the Subject

The rescuer may use the techniques described in Chapter 6 to descend or ascend to the location of the subject. When descending to a subject, the rescuer may wish to carry the descent line with him or her in a bag, feeding it out as needed. If the rescuer were to throw the rope all the way down the structure, as for descent, it could hit the subject, or the subject may be likely to reach out to try to grab the rope or the rescuer—which could destabilize either one of them, or put them in greater danger. Another hazard of throwing the descent line down ahead of the rescuer is that if the subject were to grab the end of the line and apply tension, it could impart a braking action to the rescuer's descender such that he or she might be unable to descend.

Carrying your rope in a bag while descending, as depicted in Figure 8.9, is not particularly difficult, but there are a few tricks that will help to simplify the process. First, use a rope bag that feeds rope easily. Good rope management, including carefully packing the bag beforehand to minimize twists and kinks, helps. Second, make sure that there is enough space between the top of the rope bag and your brake hand to permit good control. Finally, connect the rope to yourself in such a way that it does not hinder your movement. You can attach it to a secure point on your harness, or you may use a rope bag that attaches to your body (such as a leg bag). If the rope bag is particularly heavy, you can hang it from the descender carabiner rather than from your body. Be sure to completely open the part of the bag where the rope feeds out to help prevent snagging.

To descend to a subject who is in trouble:

1. Try to make verbal contact with the subject before descending.
2. If verbal contact is made, reassure the subject and advise him or her to hold on to whatever he or she is secured to at the moment. Tell him or her NOT to reach out for you or to try to grab you, your rope, or any of your equipment as you approach. *Note that if you must throw your rope down before descending, it is better to be offset a little further to the side of the subject so that your rope will be out of his or her reach as you descend.*
3. Establish an anchorage for your descending system almost in line with, but not directly above, the subject. This will help you avoid kicking or dropping things on him or her. However, be sure that you are close enough that you will be able to reach him or her.
4. If you plan to descend while carrying your rope in a rope bag, secure it to yourself appropriately.
5. Check your PPE to ensure that it is secure. If you plan to perform a pickoff rescue, connect your descender to yourself with two carabiners linked in line. This technique, demonstrated in Figure 8.10, will allow you and the subject to be connected to the descender independently of one another, enabling you to escape the system if necessary. *Note: While this contradicts the concept of not hard-linking multiple connectors in line, this is a case where a judicious exception is justified in that the potential disadvantages may be outweighed by the advantages, particularly since your harness connection does offer some flexibility.*
6. Determine how you will secure the subject once you reach him or her; be sure that you have the necessary equipment with you, and that it is conveniently

Figure 8.10 For pickoff rescue, connect with two carabiners in line between the descender and your harness.

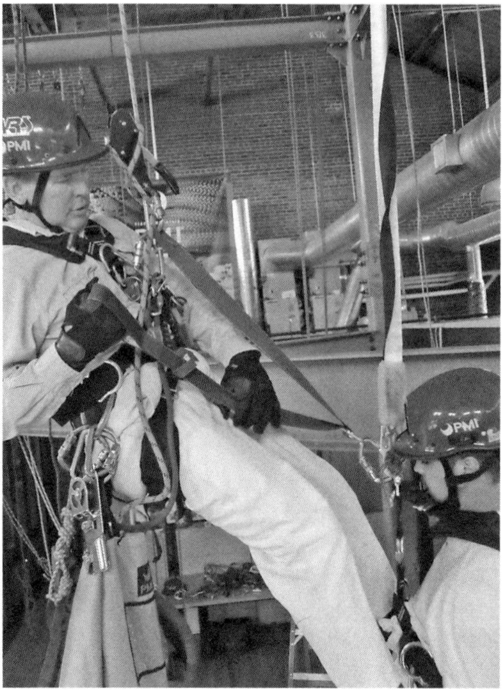

Figure 8.11 An adjustable pickoff strap provides a versatile way to connect to the subject.

accessible. One method is to use a connector to attach a "pickoff strap" to the carabiner on your descender, as shown in Figure 8.11. A pickoff strap maximizes adjustability between you and the subject.

7. Descend slowly, as described in Chapter 6.
8. When you are approximately 2 feet above the subject, come to a full stop and "tie off" your descender according to manufacturer's instructions (Fig. 8.12). Be sure to secure yourself by tying off your descender or creating a new anchorage point before attempting to render assistance to the subject. Even if the descender offers a hands-free feature, a simple tie-off can prevent you or the subject from inadvertently activating the lowering mechanism. Do not skip this step; doing so could put your life, and that of your subject, at risk.
9. While *reaching* the subject is our stated goal in this process, the step is not complete until the subject has been secured. This can be accomplished using the pickoff strap that you attached to your descender in Step 5. From your position 2 feet above the subject, extend the pickoff strap to the subject's harness attachment point. It is best to attach to the harness attachment point that is already under load, as long as this point is secure and accessible. If the subject is not wearing a harness, you may need to fit him or her with an emergency seat or other subject holding device, as shown in Chapter 7. There are several different types of emergency seats available; use yours in the manner instructed by the manufacturer. If necessary, the subject may be temporarily secured to the structure.

Figure 8.12 Rescuer should tie off his descender before attempting to connect to subject.

Once the subject is secure, manage any immediate, life threatening injuries as best as possible before proceeding to the phase of the operation that will release the subject from suspension.

Avoid connecting the subject directly to your harness, or to any component on your harness that would encumber the rescuer or that cannot be released readily. The method described above allows you to attach to the subject in such a way that will allow you to remove either yourself of the subject from the system, each without compromising the safety of the other. Do not attempt to open any connector that is under load, as this will compromise the strength of that connector.

Note that any time two hard points or connectors are attached together, you must take particular care to avoid rollout. While some may advise to never connect two hard points or connectors together, in this situation, the advantage of being able to attach and detach either rescuer or subject from the system without opening a loaded connector may justify making a calculated exception to that general statement.

Reaching the Subject by Ascent Reaching a subject by ascending poses a slightly different challenge. While the subject may already be attached to a vertical lifeline that is reachable, using this lifeline to ascend to the subject should be used only as a last resort, particularly if the subject has fallen onto it. This is because the lifeline may have been compromised by the impact. If possible, use a ladder, the structure itself, or an independent rescue line to ascend to the subject. If using a vertical lifeline or rescue line to ascend to the subject, first ensure that the line is anchored appropriately to an anchor that is sufficient for the intended (one- or two-person) load, and that it has not been compromised.

When ascending to a subject to perform a rescue, whether you are on your own line or the subject's line, ascend past the subject to a point just above the subject (approximately 2–3 feet should suffice) before connecting the subject. This will allow a certain amount of working room for your next step, which is to release the subject from the predicament and lower him or her to safety.

Ascending past the subject on his or her own line is a standard procedure for certified rope access technicians, but should not be attempted unless you are specifically trained in this method. As with all rescue methods, secure yourself before trying to assist the subject. You may secure yourself directly to the structure, or you may be secured to a vertical lifeline using appropriate equipment and techniques.

If the plan is to do a pickoff (described later in this chapter), switch over to descending mode before placing the subject on the rescue system.

Release and Extricate the Subject

Methods for release and extrication of the subject will vary slightly depending on whether you intend to descend with the subject or whether you will lower him or her from above. Because these steps are closely interrelated, the method selected for extrication will affect the method for release. For this reason we will discuss these two steps in pairs, addressing first the methods for releasing and extricating the subject by pickoff, and then addressing the methods for releasing and extricating the subject by lowering.

Releasing the Subject by Single Rescuer Pickoff If the subject has fallen to a lower level and is not suspended in his or her fall arrest system, you may not have to actually "release" him or her from anything. With a little luck, the subject may even be somewhat mobile and able to assist him- or herself in transitioning to the rescue system. Extrication may simply be a matter of loading the person into your system and lowering or descending with them.

If the person must be lifted manually to be released from suspension or to be removed from the surface on which he or she is resting, this is easily accomplished using a pre-rigged haul system such as is discussed in Chapter 7. A step-by-step guide for using a pre-rigged haul system for a pickoff is as follows:

1. First, come to a complete stop 2 feet above the subject and tie off your descender according to manufacturer's instructions for that device.
2. Connect the subject to your system using a pickoff strap, as described above. Be sure to do so in a way that allows you to remove either yourself or the subject from the system, each without compromising the other.
3. Connect a rope grab or ascender to your line, high above your own ascender, descender, or other attachment, but not quite out of reach.
4. Connect one end of a pre-rigged haul system to this rope grab.
5. Connect the other end of the haul system to the subject.

Figure 8.13 With the subject connected to your system, descend to safety.

6. Pull. With a little work, you should be able to lift the person just high enough to move him or her and, if necessary, unclip him or her from his or her original system. Be sure not to unclip the pickoff strap that connects the subject to your rescue system!
7. Once the subject is clear of the original suspension system, you may lower him or her carefully onto your rescue system (via the pickoff strap) and prepare for descent.
8. Descend to safety (Fig. 8.13). Be sure to remove the rope grab and haul system from your rope before descending.

Another option for lifting the subject from his or her predicament while on descent involves the use of a pre-engineered multifunction rescue system, such as the one shown in Figure 8.15, that integrates raising and lowering functions in an in-line system, as illustrated in Figure 8.14. With this type of system, the rescuer descends on a special type of descender that is designed to also function as part of a block-and-tackle.

1. As in the previously described procedures, descend to a position approximately 2–3 feet above the subject.
2. Connect the subject to your system using a pickoff strap.
3. Place a pulley on your rope, high above your descender, according to manufacturer's instructions.
4. Reeve the slack side of your rope through the pulley.

Figure 8.14 Use an in-line haul system to lift the subject out of his or her fall arrest system and transition him or her to the rescue system.

Figure 8.15 Pre-engineered multifunction rescue system. Credit: Heightec.

Figure 8.16 Pull on the slack side of the rope to lift the load.

5. Pull on the rope to raise first yourself, and then the subject (Fig. 8.16). As you take up slack, the subject's weight will also be lifted onto the pickoff strap. This method requires a little muscle, as you will be lifting both yourself and the subject simultaneously.
6. When you have lifted just high enough to remove the subject from his or her system, tie off the descender according to manufacturer's instructions and disconnect the subject from the original system.
7. Prepare for descent, as shown in Figure 8.17.

Although the raising method described here can be repeated multiple times to raise the subject to a high platform or other surface, the best choice is to prepare to extricate the subject to a lower level as soon as he or she is released from his or her system. This includes removing the top pulley or any extraneous system components, and following the steps outlined below.

Extricate the Subject Unless there are other rescuers available to assist (addressed in Chapter 9), extricating the subject to a lower level is usually the best and simplest choice. This may be done by descending in tandem with him or her, or by transferring the subject into the lowering system and lowering him or her from above. The pickoff method will be discussed first, and the lowering method will be discussed in the pages that follow.

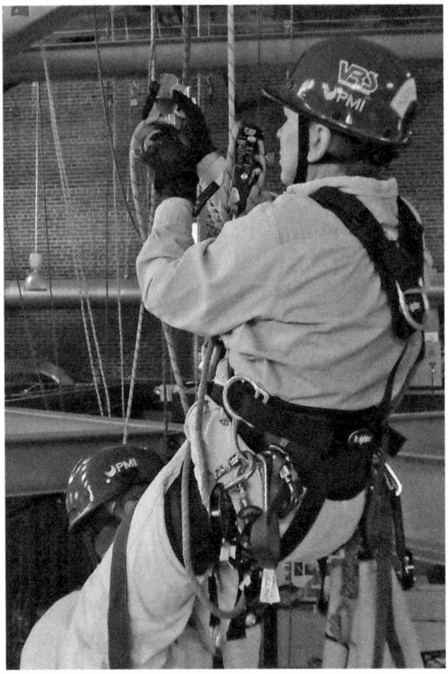

Figure 8.17 With the subject suspended from your system, release his or her original system and prepare for descent.

Before attempting a rescue pickoff, ensure that your descent device offers sufficient strength and friction to effectively manage the load of yourself plus the subject. If you have followed the steps outlined above for removing the subject from his or her system, he or she should already be suspended on your system by way of the pickoff strap, which should in turn be attached to your descending system.

Position the subject with his or her back to the structure. The most comfortable and secure position for the subject is slightly below, in front of, and facing the rescuer, with the pickoff strap running from the subject's harness up between your legs to your descender.

From this position you can manage a conscious or unconscious patient, keeping him or her clear of obstructions with your legs, while reassuring him or her verbally. Descending with a subject is quite similar to descending alone, except that the load will be heavier and you must manage and protect your subject on the way down. If you are working on a structure with beams or intermittent obstacles, such as a tower or in a plant with pipes, equipment, or machinery parts at height, you must pay special attention to avoid the subject becoming caught up in these on the way down.

To descend with a subject:

1. Check your PPE, and that of your subject, to ensure that it is secure.
2. Decide how you will manage the rope—whether you will throw it down or carry it down with you—and arrange it accordingly so that it is out of the way of the subject. If you throw your rope down ahead of you, be sure that it does not become entangled in itself or obstructions.

Figure 8.18 Position the subject for descent.

3. Arrange the subject in a position as described above, slightly below you and between your legs, so that you can effectively manage him or her on the way down (Fig. 8.18).
4. Facing the subject with one leg on either side of him or her, lean into your descending system, using your legs to direct your path of descent and to push away from the structure.
5. Descend in a slow, controlled fashion, taking care to protect the subject as much as possible on the way down. Do not bounce.
6. When you reach the ground, let the subject come to rest in a full seated position as you land on your feet. Bend your knees, or squat, so that you can descend far enough that you can release yourself from your system.
7. With the subject still seated, and still attached to the system, disconnect yourself from the system as shown in Figure 8.19. Disconnecting yourself first allows you to more effectively manage the subject, and to use the fact that he or she is still slightly suspended to help you position him or her if necessary, even while awaiting additional help.

The above methods are specific to assisted rescue techniques where the responder descends with the subject. It is also quite possible, and sometimes even preferable, for the responder to remain stationary (e.g., anchored on a platform or at a mid-point location) and to lower the subject from above. This works best where the subject is conscious and able to help manage his or her own descent.

Figure 8.19 When you have reached the ground, release yourself from the system first.

Accessing the subject is essentially the same for this type of rescue method, with the only exception being the goal of transitioning the subject to a structural anchorage rather than onto your own descending system.

Releasing the Subject for Lower from Above To lower from above, you must have access to a sufficient length of rope that will reach the ground from the location of the subject. You may either carry an extra rope with you for this purpose, or you may use the tail of the subject's lifeline, if it is of sufficient length, intact, and compatible with the rescue system. Avoid using the tail of your own lifeline or safety system, because this could preclude your ability to maneuver or escape in the event of unforeseen problems.

For the purposes of this text we will assume that the responder is carrying a complete rescue solution, including a spare rope, with him or her since this would be the preferred, more predictable, and most secure approach for a preplanned rescue. Using a pre-engineered rescue system that is capable of both raising and lowering is the best choice for extricating and evacuating a subject by lowering from above.

To lower from above, an anchor must be established at a level higher than the subject. If possible, anchor the rescue system separately from that to which you (the responder) are connected. While anchoring to a vertical lifeline is acceptable, anchoring to a structural anchorage as shown in Figure 8.20 will provide more stability and facilitate easier lowering operations.

If the subject is suspended in his or her safety system or needs to be raised to be removed from his or her position, access him or her as described above and connect him or her to your system with a pickoff strap. In this case, the strap is merely to offer additional security and keep the subject within your reach until you

Figure 8.20 Anchoring to the structure provides greater stability.

are prepared to lower him or her. It will be removed before the lower begins. Proceed as follows:

1. Connect the multi-function lowering/raising device to an appropriate anchorage that is within your reach but otherwise as high as possible above the subject.
   - The anchor should be in line with the fall line, and capable of holding at least 3,000 lbf or two times the maximum potential load, whichever is greater, in any direction that is likely to be loaded. Avoid sharp edges, and use an appropriate anchorage connector for the type of anchor chosen.
2. Connect the end of the rescue line to the subject using an appropriate connector.
   - If the subject is already suspended from a connection point on his or her harness, connect to this same place as long as it is not compromised. Using a different attachment point will make the job of raising the subject out of his or her suspended condition much more difficult.
3. Reeve the rope from the subject's harness up and through the multifunction/lowering device that is attached to the anchor, then back down toward the subject.
4. Place a pulley on the rope as near as possible to the subject's harness attachment point, or to the attachment point itself. Make sure that it is positioned in such a manner that it can be released once the subject is loaded onto the rescue line.
   - Other types of haul systems may be used here, if desired.

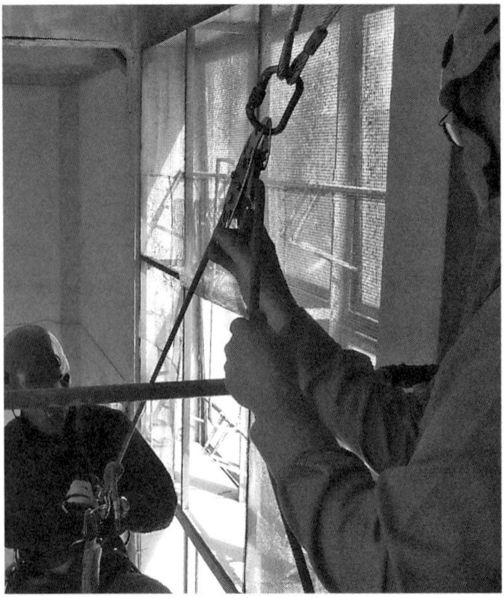

Figure 8.21 Connect the subject with a rope to the rescue system.

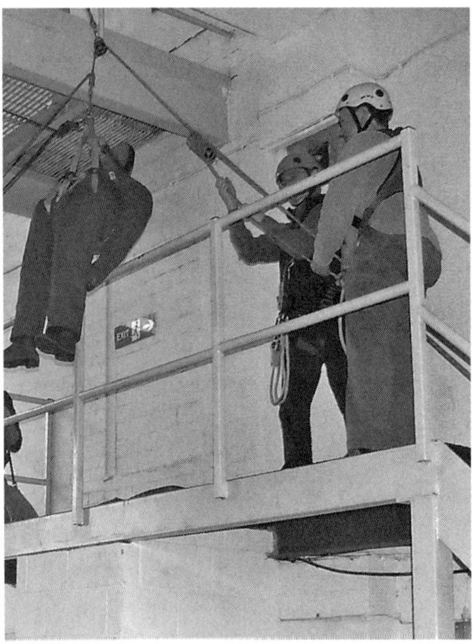

Figure 8.22 Use an inline haul system to raise the subject out of their fall protection.

5. Pull on the haul system, raising the subject just enough to release the original system. As you lift, be sure that the multi-function raising/lowering device is capturing the progress (Fig. 8.22).
   - If the direction of pull is awkward, adding a change of direction at the anchor may be helpful.

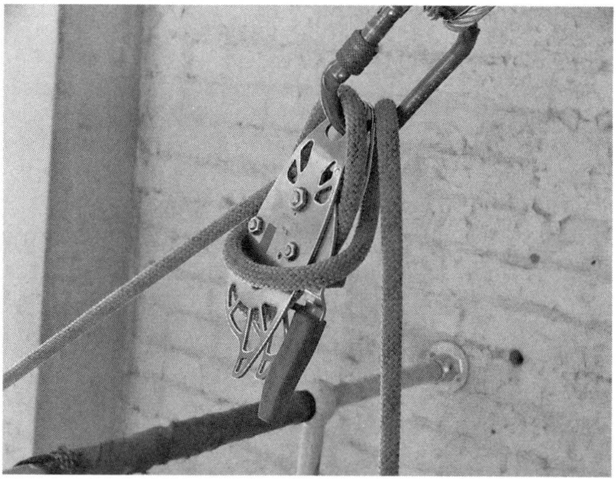

Figure 8.23 Properly secure the lowering device before letting go to remove the haul system.

6. When you have lifted just high enough to remove the subject from his or her system, tie off the descender according to manufacturer's instructions and disconnect the subject from the original system (Fig. 8.23).
    - The subject will still be attached to the pickoff strap as well as to the rescue system.

With a little luck, the subject may even be somewhat mobile and able to assist him- or herself in releasing the safety system that he or she is/was previously attached to. If the fallen person has fallen to a lower level and is not suspended in his or her fall arrest system, or if you intend to raise him or her all the way up to a higher level, you may not have to actually "release" the subject from anything.

If the situation warrants lowering the subject from their present location, it will usually be necessary to first disconnect him or her from whatever fall protection he or she was using.

Extricating the Subject by Lowering By this time you will clearly understand the advantage of using a rescue system that includes a multifunction raising/lowering device. Converting this type of system to a lower is a simple matter of removing the raising components and initiating the lower, as follows:

1. With the subject tensioned onto the mainline and the lowering device properly secured, remove the hauling component(s) (pulley) from the rescue system and convert the system to lowering mode, in accordance with the manufacturer's instructions for the device you are using.
    - If the lowering device is high above the rescuer's head, adding a connector at the anchorage to "redirect" the rope will aid in rope management.
2. Ensure that the subject is free and clear of the original fall protection system.
3. Ensure that the rope that you are using for the lowering operation is free of knots, kinks, and entanglements. Feeding it out of a rope bag is usually a good choice.

4. Disconnect the subject from the pickoff strap that is connected to your system. This strap should not be under tension, as it was there only to provide "backup safety" during the transition of the subject from the original fall protection system to the rescue system.
5. Maintaining verbal contact with the subject, first ensure that he or she is ready to descend and then, when the subject is ready, activate the lowering device to lower him or her slowly to the ground. The subject will need to use his or her feet and legs to push off of obstacles, but should not "bounce" on the system (Fig. 8.21).
6. Lower the subject until he or she is in a seated position, and ask the subject to remain there—connected to the rescue system—until you, or other rescuers, are able to reach and release him or her.

RESCUE FROM ROPE ACCESS

Rescue of a subject who fell while using fall arrest equipment differs from a rescue of a subject who was using rope access techniques. Rope access technicians receive extensive hands-on personal escape and partner rescue training, and their skills in both these areas are evaluated as part of their being tested for certification. The methods used by rope access technicians for partner rescue always maintain the two-rope system philosophy for maximum safety. The amount of training possessed by a rope access technician helps to ensure that the second rope actually does provide a safety backup and does not become entangled or pose additional liability.

A rope access technician will often choose to access the subject by ascending or descending on his or her own set of ropes as described above. However, in some cases the technician may choose to ascend the subject's rope from below. This method should only be used by trained, certified rope access technicians, and then only when the integrity of the subject's rope is sure.

A rope access technician will always be on two ropes: a primary and a secondary, whether performing work or rescue. Always ascend whichever rope the subject is not already hanging on, and use some means of fall protection (such as a rope grab on the subject's other line) as security.

Steps to co-worker assisted rescue using the subject's ropes:

1. Don an appropriate ascending system.
2. Connect your ascending system to the subject's *unloaded* rope and prepare to ascend as described in Chapter 6.
3. Connect your backup mechanism to the rope from which the subject is suspended.
4. Ascend to the same height as the subject.
5. Attach a lanyard from yourself to the subject's harness D-ring.
   - This will provide an extra measure of security to the subject throughout the operation, and will also temporarily serve as your backup as you move past him or her.

6. Assess the subject.
   - Calm the subject and treat any life-threatening injuries.
7. Detach your original backup mechanism from the rope below the subject.
8. Reattach your backup device to the rope from which the subject is suspended, but this time at a point above the subject and any connection(s) that he or she may have to that rope.
9. Ascend until you are slightly above (but still within reach of) the subject.
10. Change over to a descent system.
    - This skill is not addressed in this text. Learn it from a competent instructor before attempting.
11. Connect a pickoff strap from rescuer's descender connector and subject's harness D-ring. This will be your primary rescue connection.
12. Position an ascender above the subject and attach a haul system of your choice.
13. Raise the subject using the haul system.
    - As you raise, tension the pickoff strap until the subject is completely off the original system.
14. Once the subject is free from his or her original connection to his rope, detach it.
15. Descend with the subject, using a rope grab on the unloaded rope for backup.

The pickoff method is a skill that should be performed only by properly trained and capable persons. It is included here merely as an example to show that accessing a subject from below is feasible for a properly trained technician. For more information on becoming a certified rope access technician, visit www.sprat.org.

SUMMARY

Regardless of the rescue method used, once the subject is evacuated medical evaluation and care should be rendered even if the subject has facilitated his or her own egress from the situation and does not believe him- or herself to be injured. Transport to a medical facility may or may not be warranted, but because of the potential for internal injuries the topic should not be ignored and the subject should be encouraged to be medically evaluated. Always err on the side of caution.

There are many acceptable methods of performing co-worker assisted rescue. Those described here are simple yet versatile enough to be used in a wide variety of circumstances. For those who are interested in learning more, the methods described here will provide a good foundation from which to grow.

Although these methods are highly effective when practiced and used correctly, professional rescue should be activated immediately whenever co-worker assisted responders go into action. Co-worker assisted rescue has as its primary goals to reach the subject as quickly and safely as possible, to release the subject from the predicament, and to extricate the subject to safety. Any person who has taken a fall and/or has had to be rescued should undergo medical evaluation and assessment.

POST-CHAPTER QUIZ

Chapter 8

1. A person who has suffered a fall should receive medical attention and be evaluated by a medical professional

   a) Only if he or she is bleeding from the mouth

   b) Only if he or she can't walk after being rescued

   c) Only if he or she has had a heart attack

   d) Even if he or she does not appear to be, or does not think he or she is, badly injured

2. In the event of a fall, the first preferred method of rescue is

   a) For the subject to rescue him- or herself

   b) To call 911

   b) A professional standby rescue team

   d) Co-worker assisted rescue

3. Initial contact with the subject should ideally occur

   a) Within the golden hour

   b) By megaphone

   b) Only by professional responders

   d) Within minutes

4. A responder's first priority is the safety of

   a) The subject

   b) Him- or herself

   c) His or her co-responders

   d) The ambulance crew

5. The steps toward co-worker assisted rescue are

   a) Raise, lower, and suspend the subject

   b) Call 911, then call the subject's next of kin

   c) Reach, release, and extricate the subject

   d) Remove the subject from harm, then get them back to work

6. If possible, the preferred approach to rescue is

   a) To use entry type methods

   b) To use non-entry type methods

7. A word used to describe attaching one system to another, such as placing a pre-rigged haul system onto a lifeline, is
   - **a)** Razorback
   - **b)** Integrated
   - **c)** In-line
   - **d)** Piggyback
8. A pre-engineered, multifunction rescue system is one that
   - **a)** Is designed by the manufacturer to be used as a unit, and is capable of either raising or lowering
   - **b)** Is put together by the user from various bits of equipment that he or she has on hand
   - **c)** Can raise but not lower
   - **d)** Can lower but not raise
9. The subject may be disconnected from his or her original safety system
   - **a)** As soon as possible
   - **b)** Only when his or her safety is otherwise assured
   - **c)** Whenever you wish
   - **d)** Only with a knife
10. When descending with a subject, the most comfortable and secure position for him or her is
    - **a)** As high as possible, behind the rescuer
    - **b)** At least 3 feet below the rescuer
    - **c)** Above, behind, and facing away from the rescuer
    - **d)** Slightly below, in front of, and facing the rescuer

ANSWER KEY

1. d
2. a
3. d
4. b
5. c
6. b
7. d
8. a
9. b
10. d

CHAPTER 9

GROUP RESCUE METHODS

When you are finished with this chapter, you should understand:

- Several advantages of preparing for a group-based internal response.
- Recommended distribution of tasks for a four-person workplace response group.
- The role(s) and advantage(s) of a safety officer.
- How to select and construct an anchor for a group-rescue operation.
- How to best utilize several rescuers to safely reach, release, and extricate the subject.
- How to perform a vector lift.
- When slingshot rigging might be advantageous.
- Key aspects of a highline.
- Advantages of a traveling brake system.

The co-worker assisted rescue methods and techniques covered in the previous chapter provide an excellent foundation and offer a good solution when resources are few. However, there are also some limitations to the techniques presented there.

For example, accessing the subject using the methods described in Chapter 8 may simply be infeasible. Perhaps there is no working surface above the subject from which to work, or it could be that the area below the work area makes it infeasible to evacuate by descent; in some cases the rescuer may not have a rope that is long enough to reach the subject; alternative methods may also be required if the subject is significantly heavier than the co-worker responder. In cases such as these, a handful of trained individuals who can work together as a workplace response group to perform co-worker assisted rescue can offer some distinct advantages over a single co-worker assisted rescuer resource. Distribution of workload is among the most important of those advantages.

Clearly the rescue of a co-worker is a physically challenging task, but it is also mentally and psychologically challenging. The responders must execute the task of stabilizing and extricating the subject, all the while remaining continually aware of their own safety, adapting to changing conditions, maximizing the utilization of available resources, and formulating contingency plans. All of this, while remaining unaffected by the psychological pressure of responding to the needs of a co-worker

Falls from Height: A Guide to Rescue Planning, First Edition. Loui McCurley.

in distress—someone with whom they may work on a daily basis, and who they may know quite well on a personal level.

This chapter draws upon the reader's understanding of equipment (Chapters 4 and 5) as well as the reader's practical skills and understanding of rigging concepts (Chapters 6–8) to provide practical guidance in performing rescue with a group of co-worker assisted rescuers, functioning together within a workplace response group.

ROLES AND RESPONSIBILITIES

Distributing the workload across several co-worker assisted rescuers reduces the strain on any one person, and alleviates the physical endurance requirements placed on any one rescuer. In Chapter 3 we discussed this concept from an incident management and oversight perspective. Now, in this chapter, we will address the "boots on the ground" roles associated with the functional section identified in Chapter 3 as "Operations," and other directly related functions. Workload should be distributed based upon skills, strengths, and abilities, and responders should train together frequently enough to remain proficient in their respective roles. Aside from the obvious role of "rescuer," there are some other important roles that are best shared among available personnel.

Recommended Distribution of Tasks for a Four-Person Workplace Response Group Includes:

- **Operations Leader:** Manages tactical operations.
- **Safety Officer:** Oversees safety during rescue.
- **Primary Rescuer:** Direct contact with subject.
- **Rigging Technician:** Handles ropes and rigging; may also serve as brakeman.

If sufficient personnel are available, any or all of these positions may have assistants. Other roles may be assigned based on the needs of the operation.

Operations Leader

For starters, it is extremely valuable to appoint an *operations leader* who is a separate person from the *primary rescuer*. If necessary, the operations leader role may be performed by the same person who is serving as the incident commander (discussed in Chapter 3). The operations leader is a critical leadership role whose responsibility it is to manage all of the tactical decisions and assignments at the field level. This individual should be capable of rescue methods him- or herself, and should be well-versed in available equipment, systems, techniques, and personnel, so that he or she can make effective operational decisions. In other words, he or she should be skilled on a practical, as well as a management, level.

Safety Officer

Another critical position that should be assigned early if enough personnel are available is that of a safety officer. Of course, everyone on site should take personal interest in safety, and should not hesitate to speak up if they have a concern. However, if enough personnel are available it is helpful to have one person appointed as a safety officer whose primary role is to stand back and watch all aspects of an operation for the purpose of ensuring that hazards are mitigated and the systems used are rigged safely. The safety officer should possess a breadth of knowledge and depth of experience such that he or she is able to assess in a heartbeat whether the system is safe. The safety officer must have the expertise to make decisions about appropriateness of individual components and how they are rigged together, with consideration to how each component fits into the overall system—even if the system is rigged somewhat differently from "textbook" methods. This role is a supervisory responsibility, and can be performed simultaneously by the operations leader if personnel are limited. In any case, a rescue system should always be safety checked by someone other than the person who rigged it before it is used.

When placing multiple responsibilities on any one individual, keep in mind that increasing the number of roles that one person holds can decrease that person's ability to focus on, and to be effective in, any one of those roles.

> Appoint a safety officer to oversee safety in all aspects of an operation. The safety officer should closely monitor equipment use, anchoring, rigging belays, personal rappelling and ascending, and seat harness tie-ins. The safety officer should be among the most experienced of the team members and must be able to effectively assess safety in all aspects of the operation.

Rigging Technician

Finally, a group rescue operation will benefit from having an assigned lead *rigging technician*, who may also serve as a *brakeman* in a lowering operation. The rigging technician should be technically competent, capable of assessing a rescue site, and able to adapt quickly to changing conditions and needs. While a rescue plan should clearly outline predetermined tactics for response to known fall hazards, in practice changing conditions and unforeseen complications often dictate the need for those plans to be altered. The rigging technician must be capable of adapting to such changes as they occur.

Allowing the lead rigging technician to manage the technical aspects of the system not only frees up the primary rescuer so that he or she can focus on the subject, it also allows more versatility within the rigging itself. For example, adjusting to changing conditions or adapting to the ongoing needs of the operation (such as accommodation of the higher loads in the system after the subject's weight is added) is easier when lowering from the top. This also facilitates better rope management, since the rope is fed from the top rather than carried with the rescuer or thrown down ahead. If another rope length is required for a long evacuation, this also is more easily accomplished from the top. Finally, if the rescuer has difficulty during the rescue and must be retrieved for any reason the system can easily be converted to a raise to achieve that.

Primary Rescuer

The primary rescuer is the individual within a group effort whose job it will be to reach the subject, either verbally or physically, or both, and follow the steps to achieve rescue. If it is necessary for someone to enter the vertical environment to reach the subject, this will be the person assigned. In this case he or she will usually not descend to the subject, but will be lowered from above, as described in the following sections. This reduces the distractions on the primary rescuer, allowing him or her to focus on management of the subject and leave the raising or lowering system to the rigging technician(s). While the primary rescuer need not necessarily be the most experienced person in the group, nor be the most technically competent, he or she should be able to make good decisions on the fly and have good patient care skills. It is essential for this person to possess at least basic first aid skills but advanced medical skills are usually not necessary. The acute extrication and evacuation phases of a co-worker assisted rescue operation seldom provide opportunity to practice intensive medical interventions.

THE GROUP RESPONSE

The foundational concepts of responding to a co-worker rescue remain the same with multiple responders as with a single responder. The goal of the operation is to safely and effectively:

1. Reach and secure the subject
2. Release the subject from the predicament, and
3. Extricate the subject to a secure location.

Reaching the Subject

Even if a well-trained co-worker assisted response group is available, don't overlook the possibility of a less technical approach, such as a ladder or a bucket truck.

If a rope-based extrication is necessary, a group rescue approach is usually a better choice than a single-person approach, simply for the extra safety that it offers. Again, the best solution is always for co-worker assisted rescuers to avoid entering the vertical environment if possible. The non-entry methods discussed in Chapter 8 can be used just as well with a group of responders as with one rescuer.

Performing non-entry rescue methods with a group of responders allows personnel the luxury of separating the hands-on role from that of the person managing tactical operations, may permit a focused safety officer, and offers greater human resources and reduced workload for system operations, such as hauling.

If co-worker assisted rescuers are unable to reach the subject using non-entry methods, it may be necessary for the primary rescuer to access the subject by being either lowered or raised. When lowered from above, the primary rescuer ties into the end of the rope, which in turn is fed through a braking device that is anchored at the top of the drop. (Fig. 9.1).

Figure 9.1 Overview of rescue system. (1) Anchor, (2) raise/lower mechanism, (3) brakeman, (4) belay system, (5) belayer, (6) direction change, (7) primary rescuer.

The function of the braking mechanism is not unlike that in a descending system, but in this case the braking device remains stationary at the top of the drop, and is operated by another rescuer. The loose end of the rope can be fed directly out of a rope bag or stacked loosely near the braking device in preparation for lowering.

STACKING A ROPE FOR LOWERING

When a rope is fed directly from a coil, it is subject to twisting and kinking as a direct result of the coil. To prevent kinks and tangles while lowering, it is advantageous to "stack" the rope before initiating the lower. Stacking involves simply uncoiling the rope into a loose heap without really trying to fold or organize it neatly. The key to stacking is to start at one end and work toward the other, letting the rope simply fall where it may. The working end of the rope becomes the end at the top of the stack. Although the properly stacked rope may look a bit like spaghetti to the uninitiated eye, if the rope is fed from the top of the stack it will unstack smoothly as it is used. Keeping the stack several feet away from the brake device will help smooth the operation (Fig. 9.2).

All other things being equal, some method of lowering is usually preferred over self-controlled descent because it offers greater versatility, flexibility, and safety. While it may be possible for the primary rescuer to access the subject using the single-rescuer methods described in Chapter 8, using methods that do not require the primary rescuer to control his or her own descent frees the rescuer up to physically stabilize the subject, treat life threatening conditions, and provide psychological reassurance.

Always consider using a belay system to provide secondary backup when a lowering system is used.

Figure 9.2 Feeding a stacked rope through a brake.

The Anchorage System In selecting an anchorage for a lowering system, consider the potential load of the rescuer plus the subject, as well as the direction of pull and the possible need to convert the lower to a raise at some point(s) during the operation. Ideally, anchorages should be preselected or even installed in known locations where a rescue might be reasonably foreseeable. Competent co-worker assisted rescuers should be capable of selecting an anchor based on pre-defined options or criteria established by a Qualified Person and defined in the rescue plan.

Select an anchorage that is secure and able to withstand the anticipated loads with a comfortable margin of safety. A perfect anchorage isn't possible every time, but any anchor point used must be sufficient to the task at hand. A co-worker assisted rescue anchorage should have a strength of at least 3,000 lbs per person attached. For a manned lowering system, this usually means at least 6,000 lbs to accommodate the co-worker responder plus the subject.

The anchor system for a lowering operation should be built based on the principles discussed in Chapter 7, with consideration given to the system it will support. It should be set in line with and as close as reasonably comfortable to the subject, yet with enough working room available for the rescuer(s) at the top to work efficiently. In a lowering system the direction of pull may vary throughout the course of the rescue, as the rescuer has more ability to vary his or her route than in a rappel/descent system.

A setback of at least 4 feet from the edge to the braking device is a good goal, but of course local circumstances may dictate that this be different. It should allow for the rescuer to be comfortably positioned to operate the braking device, and for other rescuers/systems (such as a raise) to be incorporated as necessary.

Keep the anchor relatively high in relation to the edge that the primary rescuer must negotiate on descent, but not so high as to make the brakeman's job harder. If possible, a high-directional anchor may be incorporated into the system near the edge. This makes edge negotiation easier for the rescuer, while still allowing the braking device to be set back from the edge for safety.

Lowering Methods Using lowering techniques to insert a co-worker assisted rescuer permits his or her hands to remain free in order to manage the subject. This is especially important if the subject is very frightened, panicked, combative, or badly injured.

A lowering system requires that the rescuer be equipped with an appropriate harness with a waist or sternal attachment. A rope and braking device should also be selected, based on desired performance, safety factor goals, and compatibility with descender and other equipment. The braking device should have sufficient friction to manage the maximum anticipated load, but still be operable with just a single person (rescuer) load.[1]

Proceed as follows:

1. Select and establish an appropriate anchorage system, consistent with the desired fall line.
2. Stack the rope, or position the bagged rope, in a location near the anchorage that will allow it to feed smoothly through descender from the desired direction.
3. Reeve the rope through the descender, near the working end, and connect the descender to the anchorage. Tie off the descender in accordance with manufacturer's instructions, or have a brakeman control the device while the primary rescuer connects.
4. Terminate the working end of the rope and connect it to the waist or sternal attachment of the primary rescuer with an appropriate connector. He or she should be a safe distance from the edge over which he or she intends to descend.
5. Attach a belay, vertical lifeline, or other secondary backup system (if used) to the primary rescuer's harness. It is acceptable to use the same harness attachment for the backup and for the primary system.
6. Brakeman and primary rescuer confirm communications terminology, and confirm that both are ready.
7. With the primary rescuer facing the anchor, brakeman pulls the rope backward through the braking device to remove any slack, creating tension between the anchorage and the rescuer's harness.
8. Check that the area of descent is clear, and the rescuer gently leans into the system so that his or her weight is supported by the rope. At this point he or

[1]Sometimes high friction devices that are suitable for heavy loads do not function well with lighter loads, especially if they are of an auto-locking design.

she should be leaning into the system but still on a safe surface, not yet over the edge.

9. When ready, primary rescuer communicates to the brakeman that he or she is ready to descend using a command such as "down slow."
10. Brakeman feeds rope through the braking device slowly to lower the primary rescuer.
11. Primary rescuer walks backwards toward, and then over, the edge, leaning hard into the braking device as necessary.
12. As primary rescuer negotiates the edge, he or she should take care to control the path of the rope, laying it as close to the fall line as possible and padding it as necessary as shown in Figure 9.3.
13. Once over the edge, primary rescuer "sits" in the harness, legs extended forward near waist level, as during descent (Chapter 6).
14. Brakeman controls rope from the top, according to primary rescuer's commands.
15. Once the primary rescuer reaches the subject or other desired location, he or she may call for a "stop." If he or she is suspended at one location for any length of time, the brakeman should lock off the braking device according to the manufacturer's instructions. If the primary rescuer is in a safe position to release him- or herself from the system, he or she should notify the brakeman accordingly.

Perhaps the most difficult part of a lowering operation, or in fact any operation that involves multiple responders, is ensuring effective communications. When a rescuer is on rope, his or her communications must take precedence over all others. Principal communications should be between the brakeman and the rescuer, either

Figure 9.3 Protect sharp or abrasive edges as necessary.

by direct voice or by radio. Utilization of a communications relay person is permissible if conditions so dictate.

The person who is "on rope" should be the one who normally provides the cues for moving, stopping, and so on. An exception is in the case of a STOP or other urgent warning, which anyone can (and should) initiate any time it becomes necessary. One way to help avoid dangerous miscommunication is to use simple, predetermined commands during on-rope operations. It is helpful to pre-establish an agreed upon set of words that will be used between the brakeman and the rescuer during a lowering operation. The words should be selected carefully to communicate the appropriate information while altering cadence, intonation, and sound enough to prevent misunderstanding. Ideally, communications should follow a statement-and-response pattern.

One example of a good set of commands is as outlined here:

BRAKEMAN:	Ready for lower on your command! (9 syllables; last sound ". . . and!"; statement.)
RESCUER:	Am I on belay? (5 syllables, last sound ". . . lay?"; question.)
BRAKEMAN:	Your belay is on! (5 syllables, last sound ". . . on!"; statement.)
RESCUER:	Down slow! (2 syllables, last sound ". . . oh!"; statement.)
BRAKEMAN:	Lowering. (2 syllables, last sound ". . . ing!"; statement.)
EITHER PERSON:	STOP—STOP! (2 syllables, last sound ". . . op!"; statement.)
EITHER PERSON:	Stop. Stop. Why? (3 syllables, last sound ". . . why?"; question.)
EITHER PERSON:	Preparing to . . . (Varies.)
RESCUER:	I am off belay. (5 syllables, last sound ". . . lay!"; statement.)
BRAKEMAN:	Belay is off. (4 syllables, last sound ". . . off!"; statement.)

These commands may be used in altered order, as dictated by need/situation. A similar set of commands can be used during a raising operation:

BRAKEMAN:	Ready for raise on your command! (8 syllables; last sound ". . . and!"; statement.)
RESCUER:	Am I on belay? (5 syllables, last sound ". . . lay?"; question.)
BRAKEMAN:	Your belay is on! (5 syllables, last sound ". . . on!"; statement.)
RESCUER:	Up slow! (2 syllables, last sound ". . . oh!"; statement.)
BRAKEMAN:	Raising. (2 syllables, last sound ". . . ing!"; statement.)
EITHER PERSON:	STOP—STOP! (2 syllables, last sound ". . . op!"; statement.)
EITHER PERSON:	Stopping. Why? (3 syllables, last sound ". . . why?"; question.)
EITHER PERSON:	Preparing to . . . (Varies.)
RESCUER:	I am off belay. (5 syllables, last sound ". . . lay!"; statement.)
BRAKEMAN:	Belay is off. (4 syllables, last sound ". . . off!"; statement.)

Auto-locking devices, as discussed in Chapter 5, are a good choice for safety. These may feature either a panic grab mechanism that locks off if the device is

Figure 9.4 Adding friction to a braking device.

squeezed tightly, or a hands-free lock that engages if the operator lets go, or they may offer both. However, a device with variable friction may be a better choice for a lowering operation where the rescuer will be lowered to the subject with plans to then connect the subject to his or her system, because the friction required for the brakeman to effectively control the weight of two people will be greater than the friction required to control the weight of just one person. Most variable friction devices do not also offer the added security of auto-locking features; these should only be used with great care by very experienced technicians.

If a lowering device that does not feature a variable friction mechanism is being used and the brakeman finds that additional friction is needed, all is not lost. Additional friction may be added to any braking device simply by adding a low-efficiency change of direction to the system as illustrated in Figure 9.4. In fact, some braking devices require additional friction when a device is used to support heavy loads. Check manufacturer's instructions to verify how your device is to be used.

When using this approach, be sure to add the change of direction in a manner that allows the braking device to continue to function as designed and set the deviation anchor far enough away from the brake to leave the brakeman some working room. The exact distance necessary will depend on the situation and the device being used. The rope should exit the device in the direction of travel originally intended before entering the directional deviation.

To add friction to a braking device using a low-efficiency change of direction:

1. Establish a deviation anchor within arm's reach of the braking device in the direction that the standing end travels.
2. Clip a connector to the deviation anchor.

3. Pass the rope through the connector and back in the other direction.
4. Fine adjustments in friction may be made by varying the angle of the rope as it exits the connector. The tighter the angle, the greater the friction.

This method is a quick and easy way to increase friction on a loaded system.

Traveling Brake: An Alternative Approach Where the concept of a belay is desirable but personnel are limited, single-rescuer methods described in Chapter 8 may be combined with the methods described in this chapter to achieve a dual-rope system.

This technique, referred to as a *traveling brake*, requires close coordination and communication between rescuers, but when performed effectively can result in offering the benefits of both rescue by descent and rescue by lower. In a traveling brake system, the primary rescuer sets up for a rescue by descent and at the same time a brakeman at the top sets up for a rescue by lower. Both systems are operated simultaneously, with one or the other always identified as the *control brake*. Whichever system is not control brake at any given time becomes the backup/belay.

In this type of system, as with any system, primary authority for movement of the system should be allocated to the rescuer who is suspended on rope. It is this rescuer who should provide primary commands for movement of the system. The rescuer should dictate when the system moves, when it stops, and which brake is primary at any given time. Of course the stationary brakeman would have latitude to vary from this practice, usually by calling a stop, in the event of a hazardous condition or in an emergency.

An example of how commands might work in a traveling brake system might be as follows:

RESCUER: Top brake ready?

BRAKEMAN: Top brake is ready! Rescuer brake ready?

RESCUER: Rescuer brake is ready; rescuer control!

BRAKEMAN: Rescuer control. Ready on your command.

RESCUER: Down slow!

At this point, the rescuer would begin descending, while the brakeman feeds the backup line through the brake device at the top. Should the rescuer wish for the brakeman to take primary control at any time, he would simply communicate this to the brakeman, with confirmation, as follows:

RESCUER: Top control on your mark.

BRAKEMAN: Top control, now.

RESCUER: Down slow!

Here the rescuer would release tension on his or her descending system, allowing the weight of the load to be transferred to the lowering system. The rescuer would continue to allow the rope to feed through the descender as a backup, or safety, measure. When the rescuer is ready to regain primary control, he or she would communicate that desire to the brakeman as follows:

RESCUER: Rescuer control on my mark.

BRAKEMAN: Ready for rescuer control on your command.

RESCUER: Rescuer control now! Down slow.

The transition can be made quite smoothly if the brakeman simply allows the lowering line to travel through the brake a little faster than the rescuer is descending. Once the weight of the load is transferred to the rescuer's system, the brakeman would continue to allow the rope to feed freely through the lowering device as a backup, or safety, measure.

In practice, it is often desirable to utilize top control when the co-worker assisted rescuer is negotiating an edge or difficult obstruction, when managing or preparing the subject for movement, and in raising operations.

Lead Climbing In some cases, it may be necessary for a co-worker assisted rescuer to climb a structure to reach a subject. This concept is discussed in Chapter 6, but we will touch on it here also because it may indeed be considered part of a group rescue method. The term "free climbing" is sometimes misused in regulatory language to denote the concept of climbing without protection. In truth, the term free climbing comes from mountaineering terminology where it simply means to climb a structure using hands and feet, as opposed to the concept of climbing the rope itself. Technically, it is the term "free soloing" that properly refers to climbing without protection.

Terminology notwithstanding, the concept of unprotected climbing is unacceptable in the workplace. In the spirit of 100% fall protection, the co-worker assisted rescuer who finds it necessary to climb a structure should be protected by some means. One method of doing so is to trail a dynamic rope, anchoring it frequently as he or she progresses along a path.

In this type of ascent, a belayer remains in a stationary position, managing the rope by feeding it through a friction device, with minimal slack in the system but also avoiding placing tension on the lead climber. If at any point the moving rescuer should fall, the belayer locks off the friction device (which may be auto-locking) to catch the fall.

Of course, using this method to mitigate fall distances to within regulatory limits set for fall arrest may or may not be realistic, depending what regulatory requirements apply. The hazard that would be created by the user taking extra time and carrying sufficient amounts of equipment to be able to space protection so close together must be balanced against the potential risk and consequence of a serious fall. It may be determined that the consequence of a longer fall would be less problematic than the additional risk created by close spacing. This should be a conscious decision, if made, and should be noted in the rescue plan.

For the purposes of this text, we will call this method "lead climbing," that is, the rescuer is climbing beyond his or her last point of protection, but is setting anchors along the way with a trailing rope controlled by a belayer to limit his or her fall potential.

Guidance for how to lead climb is provided in Chapter 6. Lead climbing to a subject can be particularly hazardous because there is a tendency of a subject to want to reach out and grab an approaching rescuer. If the rescuer is lead climbing a structure, if possible he should climb past the subject, out of arm's reach, redirect the system through a pulley at a high anchor, and then allow him- or herself to be lowered from there. If the intent is for the rescuer and subject to be lowered to ground, this method will require enough rope to ultimately reach all the way up and through the pulley, and then back down to the ground.

Ascending Ascending a rope is also a viable means of accessing a subject, and may in fact be used in conjunction with lead climbing. If the subject is above and out of reach of responders, the first co-worker assisted rescuer may be able to ascend to the subject, and then establish a fixed line that other co-worker assisted rescuers may safely ascend. Ascending methods are addressed in Chapter 6.

Securing the Subject

Immediately upon reaching the subject, the co-worker assisted rescuer's first priority is to connect the subject to the rescue system to protect him or her from further fall hazards. If the subject is already wearing a harness, this is easily accomplished with a simple pickoff strap. This pickoff strap can be set in a ready position (Fig. 9.5) before the rescuer descends by clipping it to the rope termination where the rescuer is attached.

> *Caution!* Do not attach the subject directly to your harness. Doing so could load your harness in a dangerous manner, and will make it very difficult for you to maneuver independently from the subject, thereby creating a potential hazard for you.

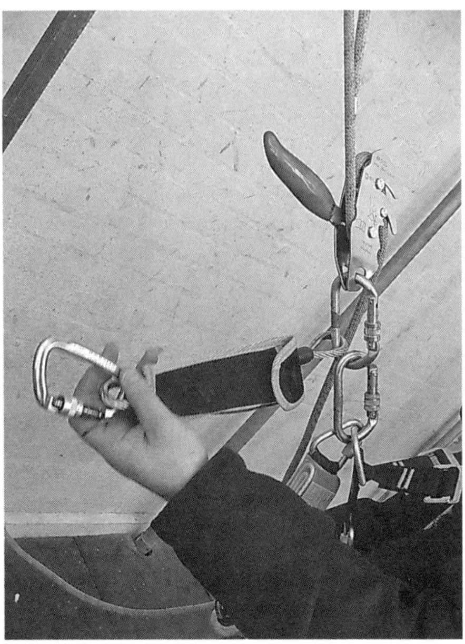

Figure 9.5 Rigging the pickoff strap in the ready position.

Figure 9.6 Connecting the subject to the pickoff strap.

This enables the rescuer to secure the subject quickly and easily simply by connecting the free end of the pickoff strap to the subject's harness attachment point as soon as he or she is within reach. One approach to this can be seen in Figure 9.6.

If the attachment point that is already under load on the subject's harness does not appear to be compromised, attaching to that, rather than to a different attachment point, will help facilitate the process of transitioning the subject to the rescue system. If, for example, the subject is suspended by a dorsal attachment, connecting to his or her sternal attachment would require you to overcome the cantilever effect created by the subject's center of gravity. The result of this would be that you would have to exert greater force and lift the subject higher in order to release him or her from the system. This transition would also be less comfortable for the subject.

If the subject is not wearing a harness at all, or if the subject's harness is compromised, it may be necessary to place him or her in a pickoff seat for the evacuation. Pickoff seats are body suspension-type devices comprised of straps, webbing, and/or solid material that are designed to be placed quickly and easily on the subject for the purpose of lifting or lowering him or her from a vertical location.

Releasing the Subject

Once the rescuer reaches and secures the subject to the rescue system, he or she must release the subject from whatever predicament the subject is in to complete the rescue. This step is more easily accomplished in a lowering operation than in a rescue by descent, because the rescuer is less encumbered and has his or her hands free to manage the subject. If possible, in a group rescue operation the majority of the rigging and lifting work should be performed by the lead rigger and other responders who are not in the hazard zone.

Some pickoff straps feature a one-way sliding buckle that allows the rescuer to tension the subject without additional mechanical advantage. If it is not possible to tension the subject enough to remove him or her from his or her suspended condition, additional mechanical advantage may be required. This can be accomplished in the same manner as described in Chapter 8, but in a group rescue operation it may be easier to let the rescuer(s) at the top do the lifting.

If only a short lift is required, one option is to perform a vector lift. A vector lift is a simple technique that involves manually redirecting the lifting line off of its path to create an angle.

To lift the subject from his or her system using a vector lift:

1. Rescuer secures the subject to the rescue system with a pickoff strap.
2. Rescuer tensions the pickoff strap as much as possible.
3. Brakeman secures the brake device by locking it off.
4. Brakeman, lead rigger, and/or other rescuers grab hold of the rescue line as near to mid-span as possible and pull down or off to one side, creating an angle. The displacement of the line will raise the load, usually just enough to disconnect the subject from his original system.
   - Those vectoring the rope *must* maintain positive control of the vectored line at all times, and not release it suddenly.

If a vector lift is not feasible, a haul system may be used. With a braking device that is designed to facilitate incorporation of a haul system, this can be readily accomplished in a matter of moments by integrating your raising system into your main line, as shown in Figure 9.7. You should have learned this skill in Chapter 8. Once the rescuer reaches the subject and connects him or her to the rescue system, the lead rigger/brakeman at the top just places a pulley on the main line and prepares to raise.

If the combined load of the rescuer and the subject is very heavy, if there is an extraordinary amount of edge friction, or if the lead rigger/brakeman is for some other reason unable to raise the load, additional mechanical advantage may be required. In this case, piggybacking a pre-rigged haul system onto the rescue line is a simple solution that will usually offer plenty of lifting capacity and distance to release the subject from his or her taut system.

The rescuer and subject should be lifted just enough for the rescuer to disengage the subject from his system in the most expedient manner possible, then the two-person load may be lowered back onto the primary system and lowering commenced.

While a raising system is arguably simple to establish and use, it is logical to ask "Why not just cut the subject's line?" The biggest reason not to use cutting of a subject's rope as a preferential method is exemplified by the infamous report of an incident that occurred when an instructor rappelled down to a participant

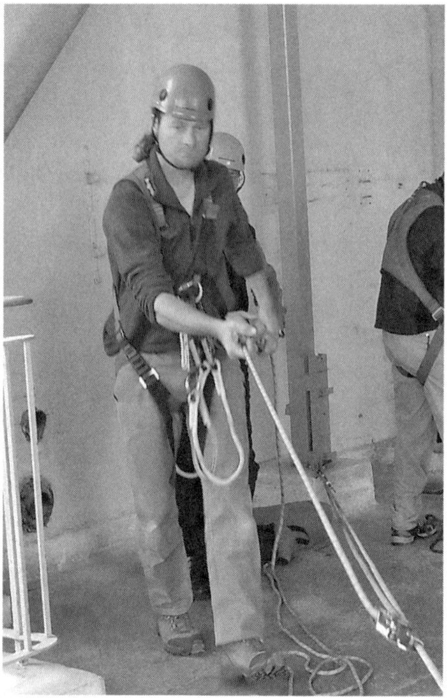

Figure 9.7 Using an integrated raising system for rescue.

whose hair was caught in her figure eight.[2] In attempting to cut her free with his knife, he accidentally severed his own rope, and plummeted to his death. While a grisly example, this incident attests to the fact that accidents can, and will, happen. Knives and flexible tension members such as rope or webbing do not mix well. The best method for releasing a subject from suspension is to take the necessary few minutes to set up a haul system and simply lift and release him or her from the system.

That said, there may be times when—for whatever reason—it is not realistic to lift a subject enough to release him or her from the system. Perhaps the subject's system is entangled in machinery, or maybe there is limited headroom. In some cases, the subject may be so heavy, or the friction in the system so great, that the available means of lifting is not sufficient to do the job. In such a case, if the subject must be removed from the system more quickly than alternative methods can reasonably be devised, cutting the subject's lanyard or vertical lifeline may be necessary.

If cutting the subject free from suspension is deemed the best choice, follow these critical guidelines:

- Use safety scissors, such as trauma shears, to help ensure that what you cut is intentional.
- Stabilize the subject; ensure that he or she is not flailing or otherwise out of control before pulling out the scissors.

[2]Jonathan Apirion, *Institutional Climbing*, 1996; http://apirion.com/ICHOME.htm.

- Be sure that the subject is connected to the rescue system and/or to a secondary/ safety system before pulling out the scissors.
- Tension the subject as much as possible onto the system that will become his or her primary support before cutting the original suspension system.
- Methodically think through, check, then recheck what will happen when the suspension line is cut.
- Nibble with the scissors, cutting a bit at a time, rather than cutting through the whole system with one giant snip.

Cutting should be used only as a last resort, and then only with great care.

Extricating the Subject

With the subject securely connected to the rescuer's system, extrication is simply a matter of lowering the load (the rescuer and the subject) to the ground, or raising the load to a higher level.

If the plan is to lower the subject and rescuer to the ground or a lower level, be sure that the subject is fully supported by the rescue system and free of his or her original suspension system before continuing the lower.

The best position for the subject to hang in is just below and facing the rescuer, and between his or her legs, as described in Chapter 8. This allows the rescuer to protect the subject using his or her feet and legs, to maintain verbal and eye contact with the subject, and to use his or her hands to help manage the subject. If the evacuation is taking place against a vertical structure, the rescuer can brace his or her feet against the structure and "walk" backward down the structure, cradling the subject between his or her legs for protection. If the operation is taking place in a free hanging configuration, the rescuer's primary objective will be to try to control spin and to watch for (and avoid) upcoming obstacles.

Steps for practicing a group-based co-worker rescue:

1. Reach the subject.
   - **a.** Connect the subject to the rescue system or to a safety system as soon as practical.
   - **b.** Use a reach pole if possible.
   - **c.** If a rescuer must enter the vertical environment, choose to lower him or her from above.
2. Release the subject from suspension.
   - **a.** Use a vector lift or a haul system to lift the subject a short distance, just enough to loosen and disconnect the suspension system.
   - **b.** If the rescuer can accomplish this easily from his or her position near the subject's location, this is acceptable. Otherwise, let the rigger(s) at the top do the work as they are in a better position to do so.
3. Extricate the rescuer and the subject by lowering or raising to a safe location.

Extrication by raise is a little more demanding on the primary rescuer, but not much. Most of the real work of overcoming gravity takes place at the top, and is performed by the rigger(s) there. Once the primary rescuer has secured the subject to his or her system and informs the brakeman of that fact, the rigger(s) simply continue to raise the load.

Raising the load a significant distance requires that a progress capture mechanism be added to the system. The term "progress capture" means just that: to capture the progress (or, amount of lift) that is achieved during the raise. How and where the progress capture is placed in the system will depend on whether the raising system is integrated into the main line (i.e., in-line), or whether it is piggybacked onto it.

In an in-line raising system, the progress capture device should be placed on the main line near the anchor on the load side of the first upward-traveling rope. In most cases this will mean that the rope grab is positioned on the main line between the first pulley at the anchor and the load.

This is true whether the system is an odd-valued mechanical advantage (e.g., 3:1), or an even-valued mechanical advantage (e.g., 2:1). Note in the even-numbered mechanical advantage example below that the purpose of the pulley at the anchor is only as a direction change, and to facilitate progress capture. Even without this pulley the system would still have a 4:1 mechanical advantage; it would just be more difficult to manage the progress capture mechanism.

Some pulleys are designed to capture the progress of a raise automatically, with an integrated camming device (Fig. 9.8). With this type of pulley, the rope can travel in only one direction through the pulley; when the rope is released the cam engages to capture the progress.

Some auto-locking braking devices are also designed to perform multiple functions, capable of being utilized at once as a lowering brake, pulley, and progress capture device (Fig. 9.9). Although these types of multifunction devices usually sacrifice some level of performance in one or more roles (e.g., pulley efficiency), system simplification and ease of operation often outweigh the disadvantages. In

Figure 9.8 Pulley with integrated progress capture. Credit: Donald Enos/SMC.

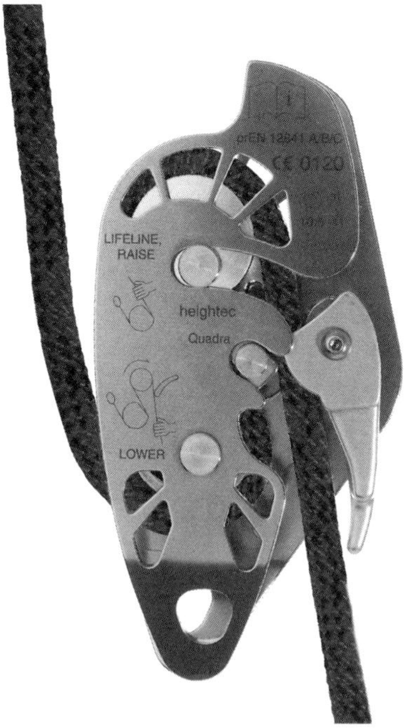

Figure 9.9 Multifunction rescue device with integrated progress capture. Credit: Heightec.

this type of system, converting from a lower to raise and back again can be accomplished in a matter of seconds.

Slingshot Rigging All of the systems rigging described above assume that the rigging is performed at a level above where the subject is located. However, all of these systems—including lowering and raising—can alternatively be set up so that the work is performed from a level *below* the subject and rescuer. To do so requires only that the rope be rigged to travel up and through a direction change above the subject, and then back down. The only caveat here is that the rope must be of sufficient length to reach twice the distance to the direction change. This type of system, shown in Figure 9.10, may be referred to as a *slingshot* system.

Although a slingshot system requires a little more rigging, more rope, and a little more effort to de-rig after the event, this type of system can be modified to include a skate block (Fig. 9.11), which provides the added potential advantage of a certain amount of horizontal movement.

All that is required to convert a slingshot system to a skate block is a lanyard, a pulley, and several connectors (Fig. 9.12). Simply attach the pulley to the line that travels between the lowering/raising device and the slingshot pulley, clip it to the subject, and lower/raise as usual.

When rigging a slingshot or skate block system, keep in mind the effect that angles have in force amplification. The force at the change of direction pulley can be twice the weight of the load.

Figure 9.10 Slingshot rigging, with operational functions on the ground.

Figure 9.11 Skate block system.

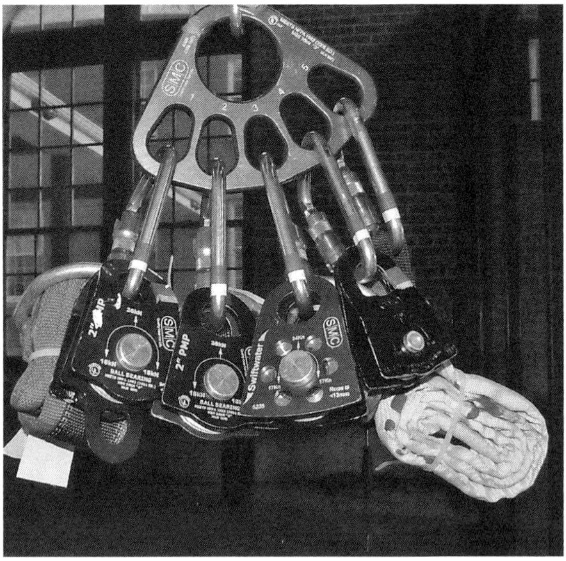

Figure 9.12 Components comprising the skate block. Credit: Pigeon Mountain Industries, Inc.

BELAYING THE RESCUER

In a co-worker assisted rescue lowering or raising operation, it is often appropriate to incorporate a secondary backup (or "rescue belay") into the system, a concept that was first introduced in Chapter 7 and then expounded upon earlier in this chapter.

In the case of a lower or raise, the term "belay" would specifically refer to the concept of a separate, independently anchored safety line connected to the rescuer and/or subject for the purpose of arresting a falling load in the event of a system failure.

Belays are considered to be standard operating procedure in professional rescue systems, especially where fail safe devices and systems are not used. In these situations, the primary suspension system is "backed up" by a secondary/belay system, the purpose of which is simply to catch a potential failure.

Preplans for co-worker assisted rescue should always take into consideration the concept of a belay. A belay is almost always warranted for training scenarios, and should be considered even in actual operations. However, the decision to utilize a belay should not be undertaken by rote, but should be considered as an important and viable *option* based on appropriate risk assessment and hazard analysis.

When it comes to safety, more does not always equal better. The decision of whether or not to belay should take into consideration both the *probability* and the *consequence* of failure. Two helmets are not better than one; two fall protection lanyards do not double your protection; two pairs of safety glasses will not decrease the probability of eye injury. Whether or not to incorporate a belay system into a rescue system is a matter of weighing the potential disadvantages against the potential advantages. Only where the potential benefits clearly outweigh the drawbacks should a belay be considered.

The first question to consider in choosing whether to incorporate a belay system is "What is the probability of failure?" In other words, what hazard(s) do you feel the need to protect against, and how likely is it that the system will fail?

Most rescue equipment and systems are rated to a static strength of around 3,000 lbs force or greater, so breakage of the system is highly unlikely unless some other extraordinary forces are at play. Other hazards, such as sharp edges, heat, and environmental contaminants can cause deterioration of system strength, but a backup/belay line would be just as susceptible to these hazards as the main line. Regardless of the hazards at play, or whether a belay is present, always rig in such a manner as to avoid potential failure.

Breakage is not the only mechanism of failure. Human failure is clearly the hazard with the highest probability of occurrence, and therefore this is the primary concern to protect against. Examples of human failure might include failure to maintain control of the lowering system or being unable to achieve rescue due to complexity or exhaustion. Getting a system "stuck" due to entanglement, twisting, jamming, or environmental contaminants would also constitute failure.

Herein lies the dilemma: in some cases the very existence of a belay system can actually increase the probability of failure due to these types of causes. Deciding to use a belay also introduces a requirement for additional personnel and equipment. Given these considerations, a belay is not always warranted in actual operations.

Again, rigging for success in the first place is of prime importance. Reducing the probability of catastrophic failure due to human error might include ensuring adequate primary system strength, using an auto-locking brake device, and using methods with which the co-worker responders are familiar and proficient. Even the decision to use group rescue methods can help reduce the likelihood of human failure because it reduces the toll on physical endurance and the amount of information that any one person must process. With all of these things in mind, jurisdictional authorities should consider whether the probability/consequence of failure justifies the use of a secondary/backup system.

Some potential advantages of belay systems include:

- May offset severe consequences of component or system failure.
- May offset severe consequences of human error.

Some potential disadvantages of belay systems include:

- More resources (personnel/gear) required.
- May increase hazard (complexity, line entanglement, etc.).

Belay systems are appropriate only when they protect against a credible threat, when it is safe to operate them, and when they prove to increase survivability for the rescuer and patient.

One way of measuring the need for a backup system is to consider the system risk in light of the *probability* and the *consequence* of failure. Weigh each of these two factors on a scale of 1 through 5 with 1 being "insignificant" and 5 being "extreme." The higher the combined total, the greater the need for a belay system. Note, however, that sometimes the combined total of *probability of failure* plus *consequence of failure* may be higher with a belay than without because a belay can so significantly increase the probability of failure.

HORIZONTAL MOVEMENT

Some circumstances dictate that a subject be moved both vertically and horizontally, either to avoid obstacles or to reach a place of safety. There are numerous options for how to achieve this, but all require greater rigging finesse and capability than straight vertical raises or lowers.

In these cases it is especially important to understand the angles and force magnification potential within rope-based systems. When rescuers begin to move a load simultaneously in multiple directions, the opportunities for force amplification increase dramatically. Only those who have received special training in horizontal systems should attempt these techniques because the methods and systems involve many angles and potentially high forces. Potential for swing fall and even catastrophic failure can be greater in such systems, but the use of belay systems further complicates the rigging, again making whether to use a belay a difficult question.

One common fall protection scenario where rescue might involve a horizontal component is in the case of horizontal lifelines used for fall protection. If a worker becomes stranded or incapacitated while on a horizontal lifeline, rescuers must first determine whether that horizontal lifeline is capable of sustaining the additional load of a rescuer and, if not, devise another approach to the rescue.

Highline Traverse

The highline traverse (Fig. 9.13) uses a stationary rope (or ropes) to connect firmly to anchors on either side of an expanse over which one might wish to cross. A rope that is secured in place across the span is called a *track line*; *dual track lines* may be used for greater strength and security. One end of the track line(s) is generally anchored in a fixed configuration, such as a tensionless tie-off, while the other end will incorporate a tensioning system to stretch the rope more tightly so that the load(s) may be moved more easily along it.

A highline traverse, used for horizontal movement of a rescuer/load, should not be confused with horizontal lifeline systems used for fall arrest. Horizontal lifeline systems for fall arrest are required to be designed to very strict specifications provided by fall protection standards. They are engineered systems, usually with sign-off by a professional engineer. Highline traverse systems, on the other hand, are intended for suspension and support, especially during a rescue, and are not intended to mimic or replicate horizontal lifeline systems in any way.

Anchorages used for highline systems must be unquestionably secure, with special consideration given to direction of pull. It is important to note that the

Figure 9.13 Highline traverse.

direction of pull against the anchors changes during use, as the load moves across the track line, deflecting it to varying degrees as it moves. Getting the track line up high can be advantageous, as the highline will inevitably sag once a load is placed on it. To help achieve height it may be advantageous to rig relatively high on vertical anchor such as a post. Especially in the case of highlines, anchors can be back tied to provide greater security.

Anchorages are often strongest at the point where they connect to the ground or other host, so another method of raising the track line while protecting the anchor is to select an anchor that is placed back from the edge, rig to it in its strongest configuration (even if that means the anchor is low), then insert a high directional such as a tripod to raise the line nearer the edge.

A highline must be tensioned for best performance. A highline that is not sufficiently tensioned will have a deep V when a load is on it, making movement of the load difficult and risking dragging the load through whatever the highline is intended to go over. On the other hand, a highline that is tensioned too tightly can impart exceedingly high forces on the anchorages and system components.

Tensioning of the highline is only performed after the fixed end anchorage is secure. The tensioning end of the highline will generally not be anchored directly, but will be connected to a haul system that is securely anchored in place. A haul system, such as that described in Chapter 7, is attached to an appropriate anchorage, and the track line is connected with rope grabs or other means, as illustrated in Figure

Figure 9.14 Tensioning end of a highline.

9.14. Note when calculating your system safety factor that this connection point may be the most vulnerable part of the system. Only rope grabs that are intended for substantial loading should be used here; most handled ascenders are not adequate for this purpose.

The highline is tensioned by firmly pulling on the haul system to draw the track line taut. Avoid over-tensioning a highline. Keep in mind the effects of vector forces in rigging, and the increased forces that will exist at the anchors and at the point where the haul system connects to the track line once the load is out in the middle of the highline. Whatever tension is applied to the track line before the load is placed on it will be amplified when the load is midspan.

If multiple track lines are used, an attempt should be made to achieve roughly equivalent tension so that they more effectively share the load. There are many possible ways to achieve this, but because such systems are used relatively seldom we will not cover rigging methods in detail in this text. Indeed, the topic of highlines is addressed only to provide the reader with an awareness and familiarity to be able to use highlines that are already rigged. The complexity of setting up highlines requires additional training and hands-on experience under the guidance of a competent instructor.

The load may be attached to the track line with a pulley or pulleys to reduce friction and facilitate movement. Highline pulleys (Fig. 9.15) often feature a very wide sheave that can roll along a single line or a bundle of lines. Pulleys of this type sometimes sport the name Kootenay. This name was originally

Figure 9.15 A highline pulley.

based on a certain region where highlines are common, but over time evolved into common usage.

Pulleys that are specifically designed for use in highlines also often feature means of attaching multiple connectors to facilitate sideways pull. This allows easy connection of tag lines, which rescuers can use to help pull the load across the highline or to slow its progress as needed. If tag lines are used, festoons should be used at regular intervals to keep the tag lines from drooping and becoming a hazard.

Festoons should be firmly fastened to the tag line in the desired position(s) and then attached to the track line more loosely so that they may travel along it without restriction. One method of achieving this is to use tied or sewn webbing or cordage loops, hitching them firmly around the tag line at regular intervals, and then clipping them with connectors to the track line as the load progresses (Fig. 9.16).

Highlines can be set up with one end higher than the other, or with both ends at roughly equivalent heights. In either case, the inherent elongation in rope will result in a certain amount of sag as the load travels across the track line.

Tensioning the highline such that the sag would be completely removed is not advisable, and this would increase the amount of force that would be imparted to the anchor(s) and/or connecting points. Instead, find ways to combat the sag or, better yet, to use the sag to your advantage.

If the load must be moved from a high point, across a void, to a lower point, the highline may be built so that it is angled from the high point to the low point. This is known as a guide line. In steeply angled systems, tag lines may need to be rigged with descending systems or haul systems to help move the load safely (Fig. 9.17).

Figure 9.16 Tag line festooned to the track line.

Figure 9.17 Steeply angled highline with separate line for lowering or raising.

Cross-Haul

Another means of moving a load horizontally is to use two (or more) hauling systems in opposition to one another. A cross-haul may be a good rigging choice where a load (subject) must be lifted up, moved horizontally, and then lowered back down again. A cross-haul, simply defined, consists of two haul systems pulling in opposition to one another. By alternately increasing and decreasing the tension on one or the other of the systems, the load may be raised, lowered, and moved from side to side.

Whereas a highline may be most appropriate when a span is wide and relatively consistent, a cross-haul offers versatility where height and obstacles are of greater concern. Rigging, and effectively implementing, cross-haul systems requires effective coordination and communication, and these systems should be utilized only by teams who have received special training and have practiced these techniques.

SUMMARY

By combining the skills and systems that we have explored thus far, any number of creative solutions can be devised to moving a subject through difficult situations. This text is not by any stretch of the imagination a comprehensive manual of all possibilities, but it does provide a solid foundation of concepts that may be exercised to achieve co-worker assisted rescue in any number of environments.

The methods and techniques discussed in these chapters require training from a competent trainer, as well as ongoing practice and retraining. Using purpose-designed, pre-rigged rescue systems as a foundation for your rescue plan will help to reduce confusion in an actual emergency, and will increase the probability of success.

POST-CHAPTER QUIZ

Chapter 9

1. If possible, assign a safety officer who
   - **a)** Is among the most experienced of available responders
   - **b)** Is among the least experienced of available responders
   - **c)** Has the most seniority among available responders
   - **d)** Is the loudest and most forceful of available responders
2. The primary goal of the co-worker assisted rescue process should be to
   - **a)** Make the employer look good
   - **b)** Wait for local emergency services responders before taking action
   - **c)** Reach, release, and extricate the subject to a secure location
   - **d)** Use as complex and complicated a system as possible to perform rescue

3. The strength of a rescue anchor should be at least ________ lbs multiplied by the number of people attached.

   a) 3,000
   b) 6,000
   c) 5,000
   d) 9,000

4. A vector lift is a simple technique that involves

   a) Two pulleys and a connector
   b) A special piece of equipment, called a vector
   c) Manually redirecting the lifting line off of its path to create an angle
   d) None of the above

5. In a system where the primary rescuer is suspended on rope and the lead rigger performs all lowering and raising functions, primary authority for initiating system operation should be deferred to

   a) The lead rigger
   b) The subject
   c) The safety officer
   d) The primary rescuer

6. A belay, consisting of a separate, independently anchored safety line connected to the rescuer and/or subject for the purpose of arresting a falling load in the event of a system failure,

   a) Should always be used during a co-worker assisted rescue operation
   b) Should never be used during a co-worker assisted rescue operation
   c) Should be used when it is safe to operate it, and where it will increase the chance of the rescuer's and patient's survival
   d) Should always be used if there is enough rope

7. A dual rope system that incorporates a simultaneous combination of rescue by descent and rescue by lower is called

   a) A traveling brake system
   b) A slingshot system
   c) A haul system
   d) A highline traverse

8. The most likely type of failure in a rescue system is

   a) Rope breakage
   b) Connector cross loading
   c) Human failure
   d) Descender failure

9. One advantage of using group rescue methods during co-worker assisted rescue is that

a) More people get to enjoy the rigging process

b) The rescuer's hands remain free in order to manage the subject

c) The employer can charge more for the rescue

d) If something bad happens, it is more than one person's fault

10. Forces at the terminating points of a highline are ___________ when the load is midspan.

a) Cut in half

b) Decreased

c) The same

d) Amplified

ANSWER KEY

1. a

2. c

3. a

4. c

5. d

6. c

7. a

8. c

9. b

10. d

PART IV

SUCCESSFUL WORKPLACE PLANNING

CHAPTER **10**

INDUSTRY-SPECIFIC CONSIDERATIONS

When you are finished with this chapter, you should have a general understanding of various workplaces where co-worker assisted rescue might be necessary. You should also be familiar with some of the unique hazards posed by different environments, including:

- Buildings and structures.
- Manufacturing and processing facilities.
- Tower construction and maintenance.
- Wind turbines.
- Agricultural facilities.
- Construction sites.
- Offshore installations.
- Bridges and support structures.
- Confined space.

As discussed throughout this book, employers are required to protect their employees from fall hazards and falling objects whenever an affected employee is at risk. The details of specific fall protection measures will vary depending on the type of work being performed.

Each unique job involves its own distinctive challenges and hazards based on the type of work being performed, the environment, and other factors. Although fall protection measures such as guardrail systems, safety net systems, personal fall arrest systems, and positioning device systems are common, the environment in which personnel are working and circumstances surrounding the job will affect how these are implemented and what rescue capabilities might be viable.

This chapter will touch upon some observations and considerations as relate to rescue for a few industries where fall hazards commonly exist. It is neither all-inclusive nor fully comprehensive. It is impossible to cover all industries or to go into great detail on any one industry in this limited space but this chapter does offer a few things to think about in context of environment and work type. The reader should use this chapter only as a starting point in identifying the kinds of things to

Falls from Height: A Guide to Rescue Planning, First Edition. Loui McCurley.

think about. It is incumbent upon the employer to ensure adequate, in-depth analysis of his or her own individual workplace to effectively mitigate hazards and facilitate rescue after a fall. This chapter does not rely on any other chapter(s) for understanding, but should be used in conjunction with other chapters to build a more well-rounded approach to rescue in the workplace.

BUILDINGS AND STRUCTURES

Workers are commonly exposed to fall hazards on buildings and similar structures during *construction* activities and while performing *maintenance* operations. *Ground-based scaffolds* and ladders are among the most common access methods used on buildings and structures.

Ground-based scaffolds are typically built from the ground up, and access is often up a ladder on the side. Structural strength may be limited, and resistance to side-loading is usually minimal. Workers on these structures are often protected by a guardrail or other means of fall prevention. More often than not the victim of a scaffold or ladder incident will fall to ground level. Although the mortality rate of this type of fall is not encouraging, at least from a rescue perspective in such cases the worker is usually relatively easy to reach, extricate, and transport.

In the event that a fall arrest or rescue lowering system becomes necessary, finding a sufficient anchor at the top of a ground-based scaffold can be a challenge. The components of most ground-based scaffolds have not been designed to accept the forces imposed in an arrested fall nor during a rescue. If rope rescue is deemed necessary from a ground-based scaffold, it is best not to use the scaffold itself as an anchorage unless the scaffold is designed with that purpose in mind and/or specific points have been approved by a Qualified Person for that application in that location.

Rigging directly to a scaffold for fall protection or rescue can cause failure of components or even the entire scaffold to overturn, which can in turn result in serious injury or death. Use only anchorage points that are approved by Qualified Persons, and consider anchoring to the structure that the scaffold is meant to access rather than directly to the scaffold. Do not overlook the possibility of rescue from above, up to and including the option of raising the subject up to a higher level, if sufficient access is available. Since the height of most ground-based scaffolds is typically within reach of ladder trucks and cherry-pickers, these are often a good choice for rescue. In considering such unconventional resources, however, keep in mind that OSHA requires load testing and use of a man basket prior to using any type of crane to lift/lower a human load.

Usually the building itself will be much more secure than scaffolds that may surround it. Nonetheless, anchorages can prove to be difficult to find and establish on buildings. While many flat rooftops feature large structures or other equipment such as air conditioning units, these are not always as solid as they look. Not being designed or intended for side-loading, this equipment is often not even bolted or secured to the building. Before attaching to anything on a building roof, consider what the original intent of that equipment was, and whether it really can withstand the sideways forces imparted by a rescue system.

Building anchors range from *eye bolts* to *davits*, *parapet clamps*, and installed anchorages. Sometimes these have even been permanently mounted into place for use by building maintenance and/or window cleaning personnel. Well-managed buildings will have a service plan document, obtainable from the building owner and/or manager, which identifies approved anchorages and their strength ratings; if possible, this document should be consulted in advance, during the rescue-planning phase.

Some older buildings have eye bolt anchors installed in or near window openings. These were originally designed for what window cleaners refer to as belt work, and should not be used for rescue or suspension.

The best choice for anchorages on buildings are those which have been identified by a qualified person as being appropriate. These often consist of beams, columns, or other structural members. Installed anchorages should be noted in the building manager's maintenance plan, and should be stamped and dated by a professional engineer. Where permanent anchorages have not been installed, consider wrapping structural parts of the building itself, such as a stairwell or elevator housing atop the building.

Suspended scaffolds, also sometimes referred to as *swing stages*, are a common sight on buildings, and have something of a reputation for being at the root of many incidents requiring rescue. A quick web search for "scaffold rescue" will reveal a plethora of incidents involving these devices. Mechanical failures, anchor failures, and wind-induced incidents result in workers left hanging from their secondary protection systems—usually a vertical lifeline with dorsal attachment—with the scaffold presenting an additional swing hazard nearby. These situations usually necessitate workers to be plucked from the sides of buildings and structures by co-workers or local emergency services, often with local news media looking on.

Most suspended scaffolds are powered by one or more motors. Some types of suspended scaffolds, such as bosun's chairs, are manually powered using block and tackle, friction devices, or other means. Bosun's chair systems are becoming increasingly obsolete in favor of more versatile systems such as rope access, but still can be found in some window cleaning applications.

Whether a bosun's chair or a larger platform-type system, rescue from suspended scaffolds is usually best performed from above, using workplace response group techniques as described in Chapter 9. Rescuers must exercise special care in these incidents so as to not become entangled in the scaffold rigging and, in windy conditions, to not be hit by the scaffold equipment.

One hazard that is particularly unique to the building environment is the potential for large numbers of people to be present during the rescue. There may be people inside, watching from windows, on balconies, on the ground below, and in other buildings nearby, and the rescuer must consider their presence from a number of perspectives.

First, the safety of rescuers is paramount. Interference from onlookers can endanger rescuers, as well as the onlookers themselves. When an incident occurs at height, the natural response is for a crowd to gather below—right in the potential path of falling objects. If possible, crowd control should be established to help keep onlookers out of access zones and hazard zones.

Another human hazard is the natural motivation of people to want to assist. Humans are a gregarious bunch, and when one is injured or in trouble, there are certain among us who are motivated to rush to their aid. The sudden influx of untrained personnel on an incident site is seldom helpful, and stands to do more harm than good. Protecting the subject, as well as protecting these well-intentioned people from themselves, becomes a concern. If people want to help, ask them to help by keeping onlookers from impinging on the rescue. Consider using police, fire, ambulance, and/or site supervisory personnel to help you enforce safety zone distance requirements.

Finally, the psychological effect on the rescuer(s) should not be ignored. A crowd of concerned onlookers who are shouting suggestions, perhaps crying or screaming with fear, watching every move with a critical eye, all the while posting continuous updates to social media, can be quite a distraction. This is not something that most people have to deal with during the course of their normal workaday lives, and it can cause the co-worker assisted rescue responder to become clumsy, forgetful, or even overly heroic. If an audience is something that you might face during a co-worker rescue, it is especially important to be extremely well-versed in the methods that you will use for the task, the limitations of the techniques that you will employ, and your protocol for ensuring that psychological pressure does not entice you to employ measures that exceed your own capabilities.

Another, less obvious, consideration in performing rescue from buildings and other structures is the potential presence of infectious contaminants found in bird excrement. Contaminants including bacteria and fungi contained in bird excrement can be quite harmful to humans, and can remain active for many months, even after the host material is dry and appears to be relatively benign. When the dried material is disturbed, it creates a dust. When the dust is breathed, the impurities are breathed into a person's lungs, and may cause illness and infection. Wearing full-coverage clothing and wearing a breathing mask will help to mitigate these hazards.

Manufacturing and Processing Facilities

The hazards that might exist in manufacturing and processing facilities are almost unlimited. Aside from typical fall hazards, in this type of environment it is typical to encounter confined spaces, machinery, hazardous materials, electrical exposures, and hazards stemming from materials used in the facilities. Preparation for rescue after a fall in a manufacturing or processing facility (Fig. 10.1) includes familiarization with the range of hazards that might exist in that facility and preparing appropriately.

Adequate preparation should include efforts to communicate with local municipal response agencies and familiarize them with your facility. Unless municipal responders know what is in your manufacturing/processing plant and how it is arranged, they will not be able to respond effectively to an incident there. A little advance planning will help responders to be safe, efficient, and successful should an external response ever be required.

Figure 10.1 Manufacturing and processing facilities can involve unique hazards.

Again, there is no substitute for advance planning when it comes to anchorages for rescue. Fall arrest anchorages are frequently installed in areas where work occurs frequently, but consider that in the event of an incident these anchorages may (will probably?) be already in use by the subject. In high-use areas where rescue is reasonably likely, consider installation of permanent anchorages expressly to accommodate rescue.

When an incident does occur in this environment, the first priority must remain the safety of the co-worker assisted rescuer. Advance planning is the best method of preventing further harm in this case, but even where a rescue preplan exists responders should evaluate the personal hazards they might encounter and mitigate those before attempting rescue. Consideration should also be given to protecting the subject from further harm during the extrication process. Standing or flowing water, mechanical hazards, high pressure systems, bulk materials, and exposed utilities are a few examples of hazards that may be present. There are also process-related hazards including thermal, chemical, and, in some cases, biological hazards that must be assessed prior to attempting rescue.

Machinery entanglement is of particular concern in manufacturing and processing facilities. Those using fall protection equipment in manufacturing and processing facilities must take special care to not allow harnesses, lanyards, and other equipment to become entrapped or tangled in the moving parts of machines.

Disentangling a person from a machine is usually a task better suited to medically trained emergency responders, for a variety of reasons. There are practical and medical considerations to take into account. Co-worker assisted rescuers will usually not be trained in the complexities of machinery extrication, nor the potential medical ramifications, and for this reason it is most often advisable to wait for trained professional emergency responders for this type of rescue. On a remote site where advanced medical capability cannot be obtained in a timely fashion, it may be worthwhile to secure a communications link with a qualified medical advisor by phone or radio during extrication by less medically qualified persons.

Regardless of who is designated to disentangle the subject, responders must ensure that the machinery itself does not continue to pose a hazard before touching it or attempting extrication. In most cases the machinery should be shut down using standard *lockout/tagout (LOTO)* procedures to prevent inadvertent restarting of the equipment during rescue. Equipment may also need to be physically stabilized before rescuers expose themselves to it. Any potential risk of moving parts or components within the machine should be mitigated, as should movement of the machine itself (rollover, shifting).

Consideration should be given to whether there are nearby utilities that might impact the rescue negatively. Any such hazards should be controlled before a rescue attempt. This may mean shutting down electrical sources (including backups), fuels, water, and gas sources. Even after utilities have been interrupted, it may be necessary to protect the rescuer(s) and subject(s) from contaminants and hazardous materials that may still be present.

If the incident involves a confined space, co-worker assisted rescue should be attempted only if the responders are adequately trained and prepared for dealing with that confined space, and the effort should be kept within the scope of their training, experience, and available equipment. Note that ventilation may be necessary before undertaking rescue even if a "permit-required confined space" is not present.

Before engaging in the rescue, establish a working area around the incident site that will prevent rescuers or others from putting themselves in harm's way or getting in the way of the operation. The area requirements will vary depending on the type of incident, number of subjects, number of rescuers, complexity of extrication, and other hazards. In general, however, there is no reason for anyone who is not directly involved in the rescue operation to be within ten yards (or more) of the site. If sufficient numbers of personnel are available to assist, establishing a manned perimeter can help prevent entry to a hazardous area.

If co-worker assisted rescuers are not appropriately trained or equipped for necessary lockout/tagout, cribbing, shoring, or stabilization methods, they should not continue the response. Instead, they should use their time to make the area as safe as possible, prevent others from entering the hazard, and wait for responders who have the necessary training, experience, and equipment.

Agricultural Facilities

Agricultural workers are exposed to a variety of hazards every day, ranging from falls to machinery to grain entrapment and engulfment. Even in large co-ops and corporately owned facilities, many of these workers work alone and in remote environments, and a deep familiarity with the environment can breed a certain amount of complacency and lack of preparedness.

Falls from ladders and catwalks onto concrete, moving machinery, grain dust explosions, and *grain entrapment* are most likely on farms, in grain trucks, and at grain handling facilities such as grain elevators. Falls are well covered in previous chapters, but it should be noted that fall protection is often deferred as neither necessary nor feasible in these environments. Fall protection should be provided for those who are working at height on trucks, railcars, grain bins, grain elevators, and on other elevated structures.

Explosions, too, are a notable hazard. Grain dust particles that fill the air can become so concentrated that they resemble a dense fog. The primary explosion often takes place inside of the grain handling equipment such as a bucket elevator or a conveyor. This primary explosion causes dust in the facility to become airborne. Once airborne, there can be an even larger secondary explosion. To help mitigate explosion hazards during rescue, use lockout/tagout procedures to keep equipment turned off, even elsewhere in the facility.

The prevalence of grain entrapment injuries makes this topic worth addressing in detail. Although awareness has grown regarding these hazards, this type of accident remains all too common among those who are involved with the loading, unloading, and movement of grain. In fact, this situation is not limited to the agricultural realm. Workers might be exposed to similar risks in any facility where bulk commodities are moved, whether grain, pelletized plastics, or other materials. Entrapment and even engulfment can occur in the blink of an eye (see Fig. 10.2).

Figure 10.2 Inside a grain handling facility, fall hazards can be compounded by potential for grain entrapment.

Incidents most often occur when bulk commodities are flowing during moving operations, or when workers are trying to release product that is stuck for some reason, such as occurs when grain becomes damp or moldy.

Access into grain handling facilities often qualifies as confined space entry due to limited access/egress and the threat of engulfment and suffocation. Flowing grain is a powerful force, whether propelled by gravity, an auger, or a conveyor system. As grain flows downhill, steeply angled slopes form in the remaining material and even a light touch to one of these slopes can cause it to avalanche. As grain moves, it takes on almost fluidlike qualities, filling empty spaces while grasping and overwhelming anything in its path. Attempts to dig a person out will usually result in their sinking deeper, with the grain filling every gap created almost before it has been vacated. The grain acts like quicksand, pulling the person deeper with every wiggle.

The dynamics of different commodities will vary. Large, variably shaped materials such as corn kernels move a little more slowly and offer more air flow beneath the surface while smaller, more consistently shaped materials such as canola can move extremely fast and pack more tightly. While it is possible for a person trapped in some grains to survive for up to a few hours even if their head is covered, breathing will become difficult for a person trapped in grain even if the airway is not obstructed due to the weight of the grain compressing the body and preventing inhalation.

Accidents can occur when a worker enters a grain bin to release grain that is stuck together or adhered to a grain bin wall in a vertical column, or when a hard, thick crust forms on the surface of the grain (also called a *grain bridge*). It is impossible for a worker looking down from the top hatch to know why grain is not moving. Should the worker enter the container and then fall through a grain bridge to the ground below, he or she may be injured or killed. Even if the worker is not injured by the fall, he or she may be buried by the resulting avalanche of grain once the bridge is dislodged.

Any time someone enters a silo, lockout/tagout procedures should be followed to disable conveyor and auger power before entry, and adequate fall protection should be provided. Even when fall protection is used, it is best for the entrant to avoid standing or walking on the surface of the grain. If it is necessary for a person to enter and walk on the grain, using a wide, flat board, a ladder, or other surface-support item to distribute the person's weight across a broader surface area may help to prevent sinking. Alternatively, the entrant may be lowered from a sturdy, high anchor point using methods that can be quickly converted to a rescue system if necessary. Rope access techniques are also a highly appropriate approach.

An appropriate rescue preplan for grain entrapment begins with having sufficient personnel on site and involved in the operation, having appropriate rescue equipment on hand, and acting quickly and efficiently.

The first course of action in any co-worker rescue incident is to notify supervisory personnel and activate local emergency services. Then, set the internal co-worker rescue plan in motion. Be aware that fumigants are commonly used in grain bins, and these can be deadly if you breathe them. While this information should be posted, due to the casual nature of many agricultural facilities this is not always the case. Do not make any assumptions; always verify the stability of an environment before entering. If the subject is trapped in a silo that has a circulatory fan, turn the fan on to get air moving. While this will not necessarily do much to clean contaminants from the air, it will help to clear the container of dust, and if the person is completely engulfed the additional airflow moving through the grain may help him or her to breathe.

Even a trapped worker who is connected to a rope safety system will probably not be able to be lifted out without first moving the grain. Depending on the commodity, even a shallow entrapment can put such force on a person that it becomes difficult to move or even to breathe. It is necessary to dig the person out before lifting to avoid trauma. Always assume that a person who has been caught in grain is alive, and initiate rescue promptly. However, only properly trained and equipped persons should attempt co-worker rescue!

Digging requires special tools. Grain tends to backfill any holes that one attempts to dig. An effective way to dig grain away from a person and to have it stay away is to use a *cofferdam*. A cofferdam (Fig. 10.3) is a temporary, self-contained enclosure that acts as a retaining wall to hold back material as it is dug out of a hole. Commercially designed cofferdams offer performance advantages, but if a commercially available cofferdam is not available, you may attempt to improvise with sheets of plywood, sheet metal, or other material. Wedge the pieces down into the grain closely together and support them from the inside to form an enclosure around the subject. The walls of the cofferdam must be supported against the compression that occurs as grain is dug out of the center of the cofferdam, otherwise the walls will collapse inward. Commercially available cofferdams offer features that will improve your chances of success.

Once the cofferdam is secured around the subject, grain may be scooped away from the subject's body and lifted out of the cofferdam enclosure using a small

Figure 10.3 A cofferdam with interlocking pieces to provide resistance against compression.

Figure 10.4 Using an avalanche probe to search for a buried subject.

bucket, grain vacuum, or whatever scoop mechanism is available. If the subject's arms are free and if he or she is physically capable of helping to dig him- or herself out, allow him or her to do so. This can provide encouragement and distraction. If the subject has been entrapped for any period of time, some medical experts recommend starting an IV before removing grain from around the subject, to help flush lactic acid from the body.

If the subject is completely submerged, or if you are unsure where in the container he or she might be, you can use an avalanche probe (or a long piece of conduit) to probe down into the commodity to feel for him or her (Fig. 10.4).

If the subject still cannot be found, or if you are unable to free him or her using normal methods, you may be tempted to cut large holes in the bin to allow grain to drain out. Consider carefully, however, the viability of this approach. First of all, this method may not be feasible for some types of silos, such as those made of concrete, and it will seldom result in a successful rescue in any case. Perhaps of greater importance, however, are the potential hazards introduced by this approach.

As mentioned previously, the dust that fills grain silos may present a combustible atmosphere in which sparks generated by a cutting tool could cause an explosion. Spark-resistant tools made of brass, bronze, copper-aluminum alloys, or copper-beryllium alloys may help, but these are often not readily at hand.

Alternatively, mitigate the explosive atmosphere (e.g., ventilate) before and during cutting, and minimize sparks by using equipment that is less likely to cause ignition (e.g., tractor bucket, air chisel, or an abrasive saw rather than a cutting torch).

Another concern with cutting into a bin is that draining the grain out of the bin will cause it to "flow." The movement of a body submerged in a flowing commodity can be difficult to predict. Depending on where the subject is located in the grain, the force of the flowing grain will likely cause him or her to move toward the hole. If the hole is not large enough for the subject to exit, he or she could become trapped against the hole while at the same time making it difficult to expand the hole without further injuring him or her.

Finally, when cutting a hole into a grain bin, consider the effect on the structural integrity of the bin itself. Cutting holes into a grain bin can create unequal pressure in the bin and can cause the sides to crease or even collapse. If a decision to cut into a bin is made, consult with a structural engineer or other qualified expert to help avoid endangering personnel.

Whatever method has been used to move grain from around the subject, once free he or she still must be removed from the container and transported to a vehicle or ambulance. The height and construction of the concrete structures at grain elevators can especially create obstacles for removing injured workers. Raising the subject out of the grain is the first challenge, and then once he or she is raised to the top of the silo, there are but a couple of options for getting him or her back to the ground.

In smaller or non-concrete silos, it may be possible to remove a subject through a side-hatch or hole, but in most cases a subject will have to be lifted through a manhole on the roof and then lowered to the ground. Do not expect people who have suffered trauma such as this to ascend and descend ladders on their own until they have received appropriate medical evaluation and clearance. Consider inclusion of lifting and lowering techniques as described in Chapter 9 as part of your overall evacuation/co-worker rescue plan.

Although many facilities have lifts, these are often quite small in size and rated only for single-person loads, making it difficult to provide patient care. In addition, some types of injuries are simply not conducive to removing a victim in the vertical position such as would be required by using the man lift. Advance planning for alternative methods of removal is imperative.

The height of a commercial grain elevator can exceed 120 feet, even just to the top of the storage silos. While there is plenty of room to construct rescue systems at this level, it is advisable to use a high directional for the edge transition. Otherwise the overhang on the top of the silo, combined with the curvature of the silo edges, makes the transition very difficult. The high directional may be facilitated through use of an artificial high directional device or through anchors on the headhouse. Avoid anchoring to equipment such as guardrails, which may not be strong enough to take the load.

Appropriate equipment for a grain elevator facility to have on hand might include an artificial high directional, retrieval system, and a cofferdam. Depending on structure height and evacuation plans, some very long ropes may be necessary (see Fig. 10.5). Consider that communications can be difficult in grain handling facilities due to noise and distances, so also be sure an appropriate communications

Figure 10.5 A method of raising the subject out of the container and then lowering him or her to ground is an important part of the overall rescue plan.

method has been established in advance, both to advise others of the incident and to facilitate communication between those involved in the operation.

Once removed from the grain, medical care for a victim of grain entrapment should include thermal protection/reheating as necessary, breathing assistance if needed, emotional support, and trauma care.

CONFINED SPACE

Many confined space accidents occur as a result of a worker (or rescuer) not even being aware that he or she is operating in a confined space. The number of confined space entrants who expire in the course of their work pales in comparison to the number of attendants and would-be rescuers who perish in an attempt to help another. Sadly, it is people who want only to help who most often fail to survive entry into a confined space. With this in mind, it should be obvious that we should pay special attention to training and preparedness of not only potential entrants but also those who might be tempted to attempt rescue, authorized or unauthorized (Fig. 10.6).

A confined space is considered to be any space that has restricted access/egress, is large enough for a person to enter/work, is not designed for continuous human occupancy, and where one of the following conditions exists:

1. Contains or has a potential to contain a hazardous atmosphere
2. Contains a material that has potential for engulfing an entrant

Figure 10.6 Only trained and competent personnel should be allowed to enter a confined space. Credit: John Peleaux, Innovative Access.

3. Has an internal configuration such that an entrant could be trapped or asphyxiated by inwardly converging walls or by a floor that slopes downward and tapers to a smaller cross-section, or
4. Contains any other recognized serious safety or health hazards (including fall, environmental, and equipment hazards).

No one should enter a confined space for any reason without first mitigating any hazards, and also planning for rescue in the event that the entry goes awry. OSHA actually calls these "permit-required" confined spaces, meaning that an employer must actually have a written plan, in advance, outlining safety and rescue procedures. OSHA has specific guidelines for developing a confined space permit, outlined in their code of federal regulations.

Fundamental to any robust confined space entry program are properly trained and qualified workers, lockout/tagout procedures for shutting down any power operations, the ability to monitor entry conditions, and the capacity to effect rescue. As discussed in previous chapters, OSHA requires rescue to be "timely" and in the case of confined spaces the concept of "timely" can take on a whole new meaning. How long can you go without air? For this reason, standby rescue teams are commonly used where confined space entry will occur.

Confined spaces are prevalent in our world, although perhaps not always readily noticed or recognized by those who have not been trained to do so. Many containers, storage bins, tanks, silos, vaults, and pits commonly found in the workplace qualify as confined spaces. Anyone who works in the vicinity of a confined space, or in any industry where one might be likely to encounter a confined space, should be trained to recognize a confined space and avoid the potential hazards that might be found there. A National Institute for Occupational Safety and Health (NIOSH) document reports that an estimated over 60% of fatalities occurred among would-be rescuers.[1] Although awareness and training has improved in recent years, would-be rescuers still become victims in an unsettling number of cases.

Although it is a difficult concept to grasp, workers should be trained to *not* enter a confined space—even if they believe the subject to be within easy reach—unless they are properly trained and equipped to do so, and are working within an existing preplan. Confined spaces often involve a combination of hazards including falls, machinery entrapment, hazardous materials, contaminated air, collapse, and more. The commitment and training to prepare for such incidents is significant. While confined spaces are an inevitable part of many work environments, the best means of maintaining rescue capability will depend on many factors. In some cases, it may not be feasible to maintain an internal response capability on an ongoing basis, but may be preferable to outsource standby entry and/or rescue teams as necessary.

There are numerous texts, instructional manuals, and consulting resources available to assist with the development of confined space entry and rescue capabilities. This text is primarily focused on the at-height and fall-related aspects of rescue and the information herein should be supplemented as necessary with information that addresses environmental hazards, such as confined spaces.

TOWER CONSTRUCTION AND MAINTENANCE

As our need for power and communications continues to grow, towers are an increasingly common sight. Guyed, self-supporting monopoles and non-standard tower structures dot the landscape in every region, soaring to heights in excess of 1,000 feet. Towers and antennae are also often coupled with other structures (such as buildings and chimneys), further complicating matters. With this trend comes an increasing need for construction and maintenance of the towers themselves, as well as the equipment on them.

Tower workers are exposed to hazards from extreme height, high voltage, microwave radiation, radio frequency exposure, icing, extreme temperatures, lightning, and gusting winds (to name a few) as they perform their respective tasks (Fig. 10.7). In addition, they frequently work with cranes, capstan hoists, drum hoists, and other means for moving heavy loads. All work and rescue activities taking place on a tower or similar structure should:

- Incorporate the use of adequate fall protection and other personal protective equipment (PPE), as appropriate.
- Observe lockout/tagout and other procedures to reduce exposure to electrical and machinery hazards.
- Equip on-tower personnel with radio frequency (RF) monitors when applicable.
- Provide for appropriate environmental protections as well as worker rehab and refreshment.

[1]DHHS (NIOSH) Publication Number 86-110, *Preventing Occupational Fatalities in Confined Spaces,* Centers for Disease Control and Prevention, 1600 Clifton Rd., Atlanta, GA, 30333.

Figure 10.7 Tower workers are exposed to many potential hazards. Credit: JP Jones, Safety LMS; original photographer unknown.

- Establish viable communications methods for workers on and off the tower.
- Have a workable rescue plan.

Work in this industry is usually accessed by climbing the tower via step bolts or fixed ladders, descending from a tower structure, or working from a manlift. In some cases, where step bolts or fixed ladders are not available or are not positioned conveniently for the work to be done, some technicians free climb the structure itself—sometimes using fall protection and sometimes without protection. It is important to note here that most cross-bracing on towers will not meet OSHA requirements for fall protection, or even rescue, requirements.

Under certain circumstances, if using a hoist that is maintained in accordance with OSHA regulations, tower erectors are permitted to hoist personnel onto and off of aerial structures during construction and removal operations. When lifting personnel using a crane, in most cases a man basket must be used. Annual tear-down of the hoist is required, and appropriate safety measures must be taken. Fall protection must be used but because fall protection is often attached to the basket itself, it is often ineffective. In addition, maintaining 100% fall protection when transitioning from the basket onto the structure can be a challenge. In truth it is quite difficult to maintain 100% fall protection when using these methods, and accidents are not uncommon.

The rapid growth of this market, high worker turnover, and narrow employer profit margins can make thorough training of workers in this industry something of a challenge. Most tower safety classes are in the neighborhood of 16 hours in duration—a time frame that most would agree is only marginally sufficient—and the knowledge and experience level of tower workers can vary widely. Tower workers should never work alone, but always in teams of at least two, and *both* workers should be fully capable of performing a co-worker rescue if necessary. There is some propensity among tower workers to leave the less experienced individual on the ground, but this philosophy can be catastrophic if the worker at the top experiences a problem.

There is no one place in construction for rules regarding tower work or construction. Each hazard has to be dealt with separately. Tower construction activities are generally regulated under OSHA CFR 29 Part 1926; however, depending upon the type of work being performed on a given tower on a given day, some parts of 1910 may also apply—specifically 1910.268 Telecommunications, which contains the fall protection requirements for communications towers, and 1910.269 Power Generation, Transmission, and Distribution, which contains fall protection and other PPE for electrical towers. Also consider ANSI A10.48, Z359, Z490, TIA/EIA 222, and several other standards. Work that involves the use of cranes or derricks falls under yet another family of OSHA regulations.

Some of the greatest challenges to rescue from towers stem from the respective heights of the towers and their potentially remote locations. This poses accessibility issues, making it difficult for on-tower workers to communicate the need for rescue, increasing the amount of time required for response, and making access to the injured worker potentially exceedingly difficult. Responders, whether co-worker or professional, may be required to use less-than-conventional methods to even get to the tower location (e.g., four-wheeler, snowcat) and once there they may need to be self-sufficient for the duration of the rescue. For workers up-tower, these parameters are exponentially compounded. Climbing a tower of any significant height is difficult work, and scooting back down to grab a jacket or take a sip of water is not really an option.

As with all work environments where falls are possible, a rescue plan should be established, documented, and communicated in advance to help improve the response to a crisis situation. The rescue plan should include all of the usual information discussed throughout this book, along with tower-specific guidelines regarding anchor points, rigging, and rescuer safety.

Tower rescue plans should include references and contact information for specialized resources applicable to tower safety and rescue, including the owner(s) of each tower, owners of equipment that is on the tower, telecommunications/RF specialists, electrical hazard/EMF specialists, trench rescue specialists, vehicle/machinery rescue specialists, tower maintenance companies/technicians, emergency incident management (overhead) teams, lightning detection equipment, helicopter short haul rescue resources, crane equipment, and tower-specific rope rescue specialists.

Ensuring appropriate response to an incident begins long before the accident ever happens. The cautionary advice in Chapter 6 bears repeating here: In the moment of a true emergency it is the most basic, ingrained reactions that take over. Most people instinctively respond emotionally to emergency situations in the workplace; this is when the training and practice pay off. When a tower incident occurs, co-workers should immediately activate local emergency services and notify company authorities according to the established preplan, and then initiate the co-worker rescue response. The co-worker responder should be protected from all potential hazards (falls, exposure, radiation, etc.) with equipment and methods that are at least equivalent to the protection that might be expected for a worker on the same tower.

Responders should be trained to examine the tower for any condition(s) that might compromise its structural integrity. Before attempting to climb the tower (and, as appropriate *during* the climb), the responder should examine the foundation of the tower and all connections, including guys, to ensure that they are secure. The stability and integrity of that structure is totally dependent on the integrity of the guying system, so this must not be overlooked. All connections, including foundation, bolted steel connections, and anchor point(s) should be considered. If suspect conditions are identified, they must be resolved prior to attempting rescue. While it may sound harsh, postponing rescue if the structure is considered unsafe *would* be the correct course of action!

Environmental hazards abound at tower sites. Responders may be exposed to such things as poisonous plants, exposure injuries (cold and heat), lightning, sunburn, and dangerous wildlife. Living things, including bats, birds, and insects, seem to be attracted to certain towers and can pose a threat from guano left on and around the tower, sudden motion that can startle a climber, and bites and stings. Special care should be taken when responding to towers in remote or wilderness locations.

The purpose of a tower should be considered in any response, and responders should prepare for associated hazards. Towers are often used to support communications equipment or electrical lines, each of which pose unique and particular hazards. Responders should not attempt to access a tower unless they know that it is safe and that hazards have been mitigated.

Radio frequency radiation is the most likely hazard on a communications tower. RF signals are usually quite directional, and the greatest hazard exists close to and in a direct line with the signal. While the comparison of RF to the effect of being cooked by a microwave oven is something of a misnomer, the hazard is real and burns are a significant risk.

RF radiation is usually transmitted from the antennae structures on a tower, but in some cases (especially where damage or leakage exists), the RF signal can travel along the tower, which in turn can act as an antenna and create exposure elsewhere on the tower. Rescuers can wear RF monitors to detect hazardous RF while they are working on or around a communications tower, but there are limitations to what the RF monitor can detect.

Communications towers also pose some electrical hazard, as the antennae on the tower are powered by electricity. FAA lighting mounted on the tower also requires electricity, and sometimes these will have a generator or big uninterrupted power supply system backing them up. Responders must identify power sources and apply lockout/tagout procedures prior to accessing the tower. Power lines usually run from a small building at the base of the tower, up a tower leg or vertical post, and then to each antenna in turn. The electrical hazards posed by these wires are usually not extreme, although short circuits and damaged wires can increase that hazard.

Power distribution and transmission towers are another story altogether. The electrical hazards that may exist on a transmission tower are many, varied, and extreme—and often the person(s) best prepared to deal with them are those who work daily in that environment. It is especially important for workers in these situations to be trained and prepared for co-worker assisted rescue, and for professional responders to consult and work alongside those with specific expertise in those fields. Never take someone else's word that a system has been de-energized and lockout/tagout applied. Someone who is knowledgeable and has authority with the tower operator should walk down the applicable systems with responders before they proceed.

The dangers of electricity are invisible to the naked eye, yet immediate and life threatening when exposure occurs. These hazards are not impossible to analyze, but doing so requires experience and training. Even a line (or other structure) that is believed to be de-energized can pose danger through residual, induced voltage, ground faults, or other situations. Rescue from an energized source or location should be attempted only in direct cooperation and consultation with the utilities involved.

Regardless of the purpose and type of tower, the preferred method of rescuing another is to do so without entering the hazardous environment. This concept of avoiding hazard holds as true with tower rescue as with other types of rescue in other environments. In the case of towers, non-entry rescue may include utilization of such resources as a crane, bucket truck, ladder truck, or aerial lift. Even these should be used with caution; always check the safety and operability of this type of equipment before attempting to use it. Special training and skills are also required for the use of these methods, and these methods may only be used where the tower and the area surrounding the tower are conducive to their use.

If a person is stranded, injured, or suspended from a high point on the tower, it may be necessary for co-worker assisted rescuers to ascend to perform the rescue. Each member of a co-worker assisted rescue team for towers should be trained at a competent climber/competent rescuer level, and should be trained specifically to identify and avoid hazards posed by the type of work and the environment at that location.

Entry-type methods for co-worker rescue in a tower environment should be utilized only when it is safe to do so and the responder(s) is adequately trained and capable of performing the required skills. If non-entry methods are not feasible and the structure must be accessed directly, a climbing ladder and/or integrated tower safety system should be used if it is available. Only responders who are specially trained and capable of doing so should attempt to access the subject by climbing the structure directly. Note that even where ladders or safety climbs do exist, it is not uncommon for climb paths to be obstructed by tower appurtenances and/or for the safety climb cable to be damaged or blocked. Responders ascending a tower for rescue purposes should have an alternate means of protection for themselves (such as a set of twin lanyards) as well as an alternate method of protection for the subject as well.

The responder(s) should climb above the subject and use methods such as those described in Chapters 8 and 9 to release the subject from fall protection and lower the subject to the ground. Responders should be prepared to deal with a subject who is resting directly on the structure, or suspended from a wire rope vertical lifeline system, fall arrest lanyard, or a self-retracting lanyard.

Tower rescue may be complicated by a need to move the subject horizontally as well as vertically. This may occur when the fall line is obstructed, for example by a widening of the structure itself, by protrusions such as antennae, or by electrical or other hazards that may exist in the vertical path. Horizontal movement is also necessitated when a subject is located on the cross arm of a tower. Methods for performing such horizontal movement are outlined in Chapter 9.

WIND TURBINES

With the rapid development of wind energy technology, serious consideration must be given by employers to the unique challenges posed for workers, and equally rescuers, at height. Because of the relatively high exposure to hazards at wind-energy production facilities, and the often remote locations (sometimes even offshore), wind farm employers and managers must develop clear roles and responsibilities and maintain a ready state for emergencies by offering simple, effective guidelines for evaluating rescue needs and making appropriate preparations in advance (Fig. 10.8).

It is the responsibility of the employer to consult with local municipal responders in advance, before a response is needed, and to assist the agency in gaining the knowledge and skills required to achieve the expected response capability. Due to the remote nature of these installations, external responders are often rural, volunteer organizations that lack the skills and equipment to pull off this kind of rescue. The owner should work with local authorities during the planning stages to facilitate their readiness and to have the equipment necessary for rescue available to them when they arrive. Even so, workers at remote sites should take special care to prepare for co-worker assisted response, and must be ready, willing, and able to respond.

Figure 10.8 Wind farm owners and managers should preplan with local municipal emergency services. Credit: Pigeon Mountain Industries, Inc.

Safety protocols within the wind energy industry can be difficult to ascertain, again because of the different types of work taking place within the context of the specialty. There are no specific OSHA rules for wind energy; however, there are a number of requirements in both Part 1926 and 1910 that apply. Each hazard would be addressed by the subpart regulating that hazard. For fall protection during construction of the turbine/tower, subpart M would apply; in general industry fall protection would be covered by several sections 1910.23 and .27, and PPE rules of .132 or 5(a)(1). As of this writing change is on the horizon as final revisions to subpart D and I rule are pending. Other standards should also be considered as applicable, including high voltage rules (NFPA 70E) in the electrical environment, and confined space and machinery guarding regulations within the nacelle.

Because turbines themselves are still a relatively new addition to our world, we are still learning much about their performance, failure modes, and associated hazards. For example, if you are attempting to pull off a rescue from a utility scale wind turbine rotor, and someone inadvertently hits the E-stop on the turbine controller (even if primary power is off) a uninterruptible power supply (UPS) system will pitch the blades to feather. If you cannot immediately comprehend the catastrophic effect this could have on a rescuer in the wrong place at the wrong time, you have just illustrated the extraordinary need for effective preplanning in this environment.

Fires are not uncommon, most of them electrical or lubrication-based in their origin. Emergency egress of personnel during a fire must be immediate, not just for protection against the fire hazard itself but for the resulting detrimental effect on air quality as well as structural integrity of the turbine. Given the heights involved and the confined space limitations, and because there is often only one way out, provision for egress must be carefully planned to quickly accommodate all potential occupants in the event of an emergency. When accessing an unfamiliar tower it is wise to be accompanied by an authorized and qualified worker for guidance in how to lockout/tagout, transition into the nacelle, location of fire extinguishers and E-stops, and locating where the emergency descent system is, what type it is, and where to rig it for use. It can be different in every turbine.

Complete fall protection must be provided for those who climb turbines. Climb distances can exceed 300 feet, and some technicians may climb several towers in a single day, sometimes carrying heavy tools. Some turbine towers are equipped with climb-assist systems which help provide a certain amount of "lift" to ease the burden on the climber. These are not typically designed to also serve as fall protection, so separate fall protection should be used. The operation of these systems is typically not intuitive, so an orientation and training on how to use them, even in the case of emergency responders, is necessary. Twin lanyards may be used where there is no preinstalled fall protection system, but cable grab systems are commonly found on the climb ladders. Ladder climbing safety systems can be rated for single or multiple climbers. This should be distinguished prior to climbing. Even where a system is rated for multiple climbers, these should be restricted to one per tower section.

In the case of very high ladders or long climb distances, rest platforms with hatches may be positioned intermittently through the climb. To protect climbers against falling objects, hatches on platforms should remain in a closed position at all times unless someone is passing through them. Platforms, whether small and proximal only to the ladder, or large and encompassing the interior of the turbine tower, are load-rated to limits of varying degrees. These load limits should be noted and strictly adhered to.

Most towers are rigged with hoists for lifting tools and machinery. Lifting is usually accommodated through openings in each of the platforms. While the openings in the platforms themselves may be protected by guardrails, falling tools and objects are a common hazard, and workers should avoid standing beneath these openings. Hoists in some of the largest turbines are actually located eternal to the tower, entering the nacelle through belly hatches. Hoists are intended for lifting tools and equipment only, and bear load-limit ratings. They should not be used to raise personnel unless they are specifically designed and approved to do so. Service lifts differ from elevators and require user training in how to operate, as well as escape from, them. You must be tied off while operating or riding in the service lifts.

Anchorages for fall protection, egress, and rescue must be identified in advance. Usually engineered/approved anchorages bear a label denoting such, and you may find that they are also painted yellow. Nacelles may be equipped externally with connection points that look a lot like they could be man anchors, but in fact in most cases these are designed and intended solely for the purpose of lifting the nacelle into place during construction, and are not rated for fall protection. The same situation may exist inside the nacelle, where lifting anchorages may be installed to facilitate equipment installation. It is possible—and even wise—to engineer and install auxiliary anchor points at strategic locations to facilitate egress or evacuation. Working with local responders to preplan response can be advantageous here.

Anchorages that are rated for fall protection or suspension should be clearly identified as such, either in the engineering drawings or on the structure itself or both. It is worth noting that many nacelles are manufactured in other countries, and may be preinstalled with fall protection anchorages that meet the criteria of that country but not others. Always ensure that the fall protection system you are using, including the anchorage, meets the regulatory criteria of the jurisdiction in which you are working.

Transition from the ladder or lift into the nacelle can be something of a challenge, and techniques differ with different kinds of turbines. For example, with one of the service lifts you exit the service lift by climbing through the roof of the lift. Then you must deploy your double lanyard and climb point to point to the yaw deck, then into the nacelle. At the nacelle itself, turbine manufacturers are increasingly requiring workers to take off their harnesses and lanyards because of the significant number of fatalities that have been caused by climbing equipment getting caught in operating machinery up-tower. Although lockout/tagout procedures should be implemented on turbines, parts may move or rotate. This can create a hazard in terms of anchorage selection as well as entanglement potential. If harnesses and lanyards must be removed for any reason, they should be left immediately available in the event that personal escape is required.

Some transformers are located at the base of the turbine they serve, while in other cases transformers are located in the nacelle. In some cases transformers and switchgear may be located on intermediate decks. In the United States, such equipment is usually enclosed, but this is not necessarily so in the case of turbines made or used elsewhere. Turbine technology is constantly evolving, and components from different manufacturers—and sometimes even different countries—are occasionally intermingled. Matters such as these should be addressed in the rescue plan to assure the safety of both internal and external responders.

As wind facilities become larger and the turbines higher, an increasing challenge is placed on construction and lifting. Within the confines of a wind facility, there is always a push to use the smallest possible crane to lift the biggest possible load, and cranes are often lifting at maximum capacity on less than ideal footing. These days, engineered crane pads are included in the turbine erection plans and related engineering drawings, and these requirements should be adhered to strictly.

A solid lift plan is essential to any crane operation, and these can be especially complex where multiple cranes are involved. Before any lift, operator credentials should be checked, crane inspections verified, and in the case of conventional cranes with lattice booms, an on-site third party inspection should be conducted prior to placing the cane into service. Where crane-suspended aerial work platforms are used, crane working load limits (WLL) should be de-rated by 50%, test lifts should be conducted, and professional engineer (PE) approvals for the work platforms and rigging should be verified.

Another method of external access to towers are blade access platforms. These are somewhat similar to the platforms used for window cleaning, featuring motorized traction hoists that move the platform up a tower or around a blade from the tip to the root for inspection or service. These require a good bit of technical rigging and operators must be specially trained. Fall protection is required while on the platform and an escape system must also be available.

Rope access is one method that is quickly becoming a preferred method for external access to wind turbines. Rope access is extremely versatile, less affected by winds and terrain, reduces fall potential dramatically, and is more efficient in terms of productivity than many other alternatives. Rope access methods are backed by a proven safety record, and certified rope access technicians are generally quite well trained in partner rescue, "just in case."

With wind installations just now reaching an age where wear and tear are proving a concern, incidents resulting from tower vibrations and fatigue in the structures are beginning to reveal themselves. Further, the drive to produce often means that construction activities are taking place on a 24/7 schedule with personnel who may not be all that familiar with these types of structures. It has been reported that an estimated 70% of the labor hired is usually local and not familiar with wind towers.

Rescue efforts may be hampered by the fact that wind farms may be large, with great distances between turbines. They are also often far from big city resources, and local municipal responders may be of a volunteer nature and neither equipped nor trained for potential incidents. Few people—whether workers or rescuers—are trained in turbine evacuation and rescue, and the unknowns are great.

It is imperative that any technicians working on turbines be trained and capable of at least self-rescue, and preferably partner rescue as well. In order for it to be effective, egress and rescue equipment must be available within reach of anyone who is working in a nacelle. Due to the potentially long time to rescue, it is also advisable to keep first aid kits and AEDs reasonably accessible (either in towers or trucks) as well.

BRIDGES AND SUPPORT STRUCTURES

Rescue from bridges may become necessary when maintenance workers, inspectors, or even the general public access portions of a bridge for any reason. Effective rescue may require close cooperation between those who best know bridges and those who best know rescue, working together. How the bridge is constructed, where a subject is located on the bridge, and the resources available will together dictate the safest and most expeditious means of rescue.

Access to persons who may have fallen from any bridge structure can be difficult, and should be considered in advance based on the potential for an incident, risks involved, and other personnel available. For example, many times the work underway on older bridges involves the stripping away of old lead-based paint. The potential health hazard presented by these situations (containment structures) will complicate the approach to rescue.

The approach to bridge rescue will depend greatly on the type of incident, what type of bridge is involved, available anchorages, accessibility, and skill of the responders. Seldom will a bridge structure offer a simple access zone, and usually the reason that the bridge exists (vehicle traffic, foot traffic, water, obstructions) will create significant limitations to a rescue incident (Fig. 10.9). It may be necessary for the rescuer(s) to move in an upward, downward, horizontal, or angled plane in order to avoid obstacles, reach the subject, and then extricate him or her. Due to the wide variations in type, construction, and environment, rescue plans must be quite site- and bridge-specific, even on structures of similar size and type.

On a positive note, at most bridge work sites there is usually a crane, under-bridge vehicle, or bucket truck present that can be implemented into the plan. For sites that don't have this, then a more personnel-intensive plan will be needed. Usually the best choice will be to lower an incapacitated subject to a safe location without any rescuers having to enter the vertical hazard, if this is possible. Raising and horizontal movement are much more difficult to achieve, and often by the time a complicated system might be rigged to raise a person it would be quicker to have emergency responders close lanes of traffic below and simply lower the person to the responders.

Rescue by lower obviously requires preplanning and rigging in advance for an event that you hope doesn't happen, and is really only feasible in the case of a worker who is able to use appropriate equipment. If lowering the subject without entering the environment is not feasible, the next best option is for the rescuer(s) to try to find a way to work from a point above the subject. Bridges tend to be by design very sturdy structures, so there is usually a plethora of good, structural steel anchorages quite handy. Co-worker rescuers can use climbing methods defined in

Figure 10.9 Rescue methods on a bridge will be at least in part dictated by the type of bridge and the location of the subject. Credit: Burgess & Niple.

their work plan to access a high point above the subject from which to work. However, just because a rescuer is able to access a location above the subject does not mean that it will be possible to lower the subject straight down. Especially in the case of a bridge over water, a roadway, or other obstructions, deviations may be required to clear hazards.

Accessing a high point from which to initiate a rescue is easier in some cases than others. Depending on the situation, there may be a platform available above the level of potential rescue. In the best of circumstances, platforms may exist in the form of roadways, footpaths, or catwalks that are designed into some bridges specifically for maintenance purposes. Where possible, using a platform that is intended for human occupancy is usually a good choice.

That said, reality is that many bridge structures are not built with post-construction access in mind, and in fact there may *not* be readily accessible platforms that can be used. In other cases, the person in need of rescue may be above the bridge platform, for example suspended from fall protection on a flyover or overpass. In these cases it may be necessary to access a beam or some other structural portion of the bridge that is not normally inhabited by personnel. Access may be gained by climbing the structure itself, but only individuals specifically trained in bridge access should attempt such access, if indeed possible. In addition to anchor consideration, bridge workers must be well versed in protection against pendulum, as well as in working above possibly inhabited roadways and walkways.

In most cases, the trusses themselves, guardrail systems, and most beam connections will suffice for anchor use—although in the case of pedestrian bridges, historical bridges, or damaged bridges, additional anchor considerations may be necessary. Responders must be very careful to avoid anchoring to conduit or other utility lines and connections. These are often not designed to accommodate large loads. When possible, bridge design drawings and/or a structural engineer should be consulted in advance, and potential anchor points identified as part of the rescue plan. Suspension, cable stay, and concrete bridges can prove to be more challenging to access than other bridges because the cables and concrete used are often not conducive to climbing. For work involving these types of bridges a site-specific access and rescue plan should address these considerations.

Some suspension bridges have fall protection cables running alongside the main cables, but these are often old, deteriorated, and of questionable condition. It is best not to rely on them at all, and especially if the fallen worker is suspended from a questionable pre-installed fall protection cable, co-worker assisted rescuers should try to use a different anchoring method. Utilization of lead climbing techniques described in Chapters 6 and 9 provides a solid foundation for approaching this problem. When lead climbing in a horizontal direction, intermittent anchorages should be placed to ensure that the climber/rescuer does not endure a catastrophic pendulum in the event of a fall. This can be a daunting exercise, however, as the distances and amount of rope length required for belaying can be significant, a situation potentially compounded by wind and physical exertion. The time, effort, and practical application of these techniques should be compared with the use of a crane, bucket truck, or other traditional access method to determine the best approach.

In the case of bridge incidents it is quite likely that there may also be a horizontal component involved in a rescue. It is not always feasible to position oneself directly above the subject, so alternative methods may need to be implemented to approach from two opposing points. See the discussion on cross-hauls and guide lines in Chapter 9 for more information on these techniques. Always keep in mind potential sag in horizontal systems, and how any traffic, obstructions, and the like below will affect this.

Where workers are suspended from beneath a bridge deck, rescue can be especially challenging. Again, the best choice is to lower the subject; however, this may not be feasible if the feature that the bridge crosses poses additional hazard, or would then be inaccessible. Many bridgework employers and bridge owners require a rescue boat for over-water work, and lowering to such a craft is indeed a viable solution. However, this is not always available or feasible.

If the ground beneath the structure is unsuitable for evacuation, utilization of raising techniques described in previous chapters may be required. Again, especially in the case of bridge work, cross-haul and guide-line techniques may prove especially useful.

Where bridge work is frequent or extensive enough to merit the costs of installation, safety nets are sometimes used for fall protection. Rescue from safety nets is

Figure 10.10 Rescue from this location would be difficult. Credit: Burgess & Niple.

discussed in Chapter 9. Safety nets pose additional challenges in that they may only be used where there is sufficient clearance, and additional exposure occurs during installation. Even the use of safety nets can increase a worker's exposure, simply due to the fact that safety nets must be inspected at least weekly and debris must be removed from nets daily. This means that there is relatively frequent access to the nets, whether intentional or unintentional, thereby increasing the likelihood of an incident.

Not all bridges extend over water, but for those that do OSHA rules for construction work over water may be found in 1926.106. Historically, there has been a tendency among regulatory authorities to require that workers over rivers and waterways employ a *personal flotation device (PFD)* in case of a fall. However, the bulk and restrictiveness of a PFD are more likely to impede other equipment, get in the way of the worker, and reduce safety overall rather than increasing safety. It has also been shown that, in the event of a fall from high above water, a PFD is likely to be stripped from the wearer as soon as he or she hits the water. For these reasons, PFD requirements for people using rope access or fall protection on bridges are often not a reasonable expectation.

Finally, it is worth mentioning that bridges are notoriously popular among emotionally disturbed individuals who wish to express discontentment and/or terminate their lives. Even if the subject is a co-worker, this is a matter best left to emergency services personnel who are trained in both the psychological and practical aspects of such situations.

Bridges can be complex and diverse structures, often difficult to access. Especially in the case of bridges that are located in rural areas, local responders may not have to the have technical expertise required to perform rescue. It is especially important in such situations for the employer to preplan effectively and ensure that personnel are appropriately trained to resolve any potential incidents that are likely to occur.

CONSTRUCTION SITES

Falls injure and kill a great number of workers in the construction industry, usually as a result of workers falling off ladders, scaffolding, and unguarded building edges and through roof/floor openings or skylights. Many of the falls that occur are from relatively low levels (i.e., under 20 feet) but can nonetheless be fatal.

The term "construction" is such a broad, far-reaching one that the meaning is almost limitless. Construction occurs on every type of structure that has already been discussed in this chapter, including buildings, towers, wind turbines, grain handling facilities, monuments, bridges, and dams. The concerns and considerations for each of these specific environments should be taken into consideration in the case of post-fall rescue from a construction incident, along with considerations related directly to the construction environment.

Construction sites often pose especially complicated rescue scenarios because of the constantly evolving environment. Hazards come and go, as do anchor points, personnel, and work plans. Accidents common at work sites include ground-based scaffold and ladder incidents, falls from height, trench and excavation incidents, and electrocution. Workers may be protected by positioning systems, fall arrest, guardrail systems, fences, barricades, or covers. Construction workers are often exposed to more than one hazard at a time, further complicating the potential rescue scenario.

Due to the high number of incidents noted in construction, particularly at height, it is not unreasonable for host employers to maintain a skeptical attitude and require the contractors who will be working at height to prove their competency and that they have a comprehensive managed fall protection plan—including a rescue preplan—in place. Even then, it is wise to verify the serviceability of equipment and to monitor ongoing performance.

The safety of the rescuer is paramount, as always. Anyone who might be called to perform co-worker assisted rescue while working in a construction site should have a though understanding of the hazards present at that site and the methods being used to mitigate those hazards. Specifically, those who might potentially become co-worker assisted rescuers must have a firm grasp on the nature of the fall hazards that exist, all fall protection systems that are being used on site, and how each of those fall protection systems should be properly implemented and used.

Co-worker assisted rescuers should also be aware of any limitations to access that have been designated on the site, and what role each employee and subcontractor has with regard to his or her own authority as well as the safety of others.

Additional hazards that must be considered in a construction environment are usually related to the job being performed. There are representatives of many trades working on a construction site, including laborers, carpenters, roofers, ironworkers, and painters. The work that the subject was performing when the incident occurred may have involved chemicals, power tools, heavy machinery, welding equipment, or other hazards that may become a factor during the rescue. It may be necessary to disconnect power and/or mitigate other hazards before initiating a rescue attempt.

Sadly, most falls in a construction environment are to ground level, and do not require complicated rescue. On a more positive note, there are often many resources available on a construction site that can help facilitate rescue, especially from the low heights where incidents often occur. Do not overlook the possibility of using a ladder or man-lift if one is available.

OFFSHORE INSTALLATIONS

Offshore installations in industries such as oil, gas, wind, or shipping can make fall protection—and subsequent rescue—extremely challenging. When you combine the "normal" hazards presented by any one of these workplaces with the marine environment issues, you really have a full plate of hazards to consider.

Floating and tethered structures, in particular, offer increased stability issues when one is accessing and working upon them. Contours of offshore structures are notoriously complex, involving harsh environmental conditions; steep, slippery climbing structures; and a relative dearth of engineered anchorages. Geometry of access paths and working surfaces can range from vertical to horizontal to steeply angled in any direction. Coated with salt, ice, and machinery grime, the worker in these environments must work on such surfaces while being pounded by wind, driving rain, or even freezing snow, all the while contending with the psychological effects of the continuous, relentless wave action of the sea below.

Workers access the offshore installation either from above, by helicopter, or from below, by boat. Due to the hazards associated with these access methods, many contractors require workers to go through specific safety and emergency training that includes practicing escaping a helicopter passenger frame that is inverted and dropped into a tank.

When accessing the outside of a marine platform from a boat or helicopter, as if fall protection harnesses and equipment are not cumbersome enough, the worker may be required to wear a full-coverage exposure suit that offers environmental protection and floatation in case he or she accidentally drops into the water. Accessing such a structure from a boat is especially challenging, as the worker must step from the pitching and rolling boat to the stable platform without a slip.

Many offshore structures do not have ladder safety systems installed, or if they do exist they may be corroded or otherwise rendered inoperable. Ladders and working surfaces are often offset and not conducive to installation of engineered systems. Although the recommended procedure in such a case would be for the worker to use twin lanyards or some other method of climbing protection, in truth these methods are often deemed too cumbersome or inconvenient and are not utilized.

A worker in an offshore environment should be prepared to deal with difficult fall protection needs, a shortage of anchorages, harsh environment, and confined spaces. Co-worker assisted rescue, then, involves the same. Co-worker rescue may be required for a worker suspended in fall arrest, a worker who has been injured, a medical emergency, or one with a heat or cold emergency. At sea, a simple evacuation by descent is seldom the best option, since what is below is the sea. However, most offshore installations do have support boats patrolling the area, and these may be utilized as a platform to which to evacuate. In most cases, the methods described in Chapters 8 and 9 will provide sufficient skills for post-fall rescue, but some of the details must be adapted to the location and hazards specific to the site.

Anyone working on an offshore facility must be trained in survival methods applicable to the situation in which he or she is working, emergencies that are likely to occur, and the offshore environment. This may include such things as operating in a confined space, chemical emergencies, how to extinguish a small fire, specific first aid skills, lifeboat operation, and survival at sea.

SUMMARY

Clearly the most difficult parts of any rescue are created by the environment and circumstances in which the rescue takes place. Medical care, ropes and rigging, and transporting an injured person are all simple concepts that are relatively easy to grasp. The real challenges are created by the hazards that are inherent in the environment, the type of work being performed, the circumstances specific to the rescue, and other location-specific factors.

It is the assumption of this manual that those reading it are already experienced and knowledgeable in their areas of employment and the environments where they work. This chapter is offered merely as a reminder that it is your specific knowledge of your workplace environment that makes you most valuable as a responder in your workplace, and to remind you that other environments offer additional challenges.

Medical considerations are not thoroughly addressed in this book, but attention should be given to this matter. As discussed in previous chapters, coordination with local emergency services organizations is key to success in the event of an actual emergency, and these can also be an invaluable source of medical care. Sometimes a worker who has been rescued is overly anxious to get back to work, either because he or she doesn't feel that he or she was really compromised, or perhaps may want to prove to his or her commitment to his employer. *Any* time a worker has been involved in a workplace incident, he or she should be encouraged to seek medical attention, if only for observation, and to get medical clearance before returning to work.

Hopefully this chapter, along with other chapters, has reminded the employer that he or she should take initiative to familiarize local responders with the workplace and the particular hazards there. Working closely with your local emergency services in advance, sharing your emergency response plan, offering training opportunities within your facility, and even helping to equip your jurisdictional responders with specialized equipment will pay dividends in results in the event of an actual incident.

POST-CHAPTER QUIZ

Chapter 10

1. The importance of understanding different workplaces where co-worker assisted rescue at height may be necessary is

   a) Minimal, as hazards remain the same no matter where you are

   b) That regulatory requirements differ, depending on location and type of work

   c) That hazards can vary depending on the workplace

   d) Answers b and c are correct

2. True or False: You can always find out where the approved anchorages exist on an urban building by consulting the blueprints or building plan.

3. A bosun's chair

   a) Is regulated under suspended scaffold rules

   b) Is the same thing as rope access

   c) Is a type of ground-based scaffold

   d) Answers a and b are correct

4. In addition to at-height risks, hazards in manufacturing facilities may include

   a) Confined spaces

   b) Machinery

   c) Hazardous materials

   d) All of the above

5. Co-worker assisted rescue in agricultural facilities

   a) Is not required because many of these are family owned and operated

   b) Can be hampered by multiple hazards, slow municipal response times, and a certain amount of complacency

   c) Is only necessary if there is a potential for grain entrapment

   d) Is only required when OSHA demands it

6. Some of the hazards that co-worker assisted rescuers should be prepared to contend with in the tower environment include

   a) RF exposure

   b) Electrical hazards

   c) Lightning and extreme weather

   d) All of the above

7. The nacelle, or tops of wind turbines,

   a) Can only be accessed from the inside

   b) Can only be accessed from the outside

   c) May be access from inside or outside

   d) Can only be accessed by helicopter

8. Co-worker assisted rescuers in offshore installations should be prepared to contend with

    a) Difficult fall protection needs

    b) A shortage of anchorages

    c) Harsh environment

    d) Confined spaces

    e) All of the above

9. If a co-worker assisted rescue responder encounters hazards during a response for which he or she is not trained, skilled, and comfortable, he or she should

    a) Respond anyway, and do his or her best

    b) Consult a book for guidance before he or she proceeds

    c) Ask someone for coaching

    d) Wait for qualified assistance

10. True or False: As long as an individual understands fall protection, ropes, rigging, and lowering/raising systems, he or she is qualified to perform co-worker assisted rescue in any environment he or she might encounter.

ANSWER KEY

1. d
2. False
3. a
4. d
5. b
6. d
7. c
8. e
9. d
10. False

CHAPTER 11

PRACTICAL WORKPLACE APPLICATION: YOUR STEP-BY-STEP GUIDE TO DEVELOPING A CO-WORKER ASSISTED RESCUE PLAN

This chapter is the "workbook" component of this book. It contains forms and worksheets that will help the employer, safety manager, or rescue leader incorporate all of what you have learned in this book into a single, comprehensive plan.

By following the steps in this chapter, you will have prepared a complete co-worker assisted rescue plan, including:

- Understanding why you need a comprehensive managed fall protection plan.
- Analyzing your plan in light of a fall hazard assessment.
- Developing a policy statement.
- Developing training guidelines for co-worker assisted rescue responders.
- Assigning specific roles and responsibilities for program administration.
- Establishing a working relationship with local municipal agencies.
- Creating a recordkeeping system.
- Implementing a plan for analyzing incidents and near misses.
- Auditing a program to ensure readiness.

Clearly, co-worker assisted rescue requires a significant commitment on the part of the employer as well as those assigned to the task. The task need not be overwhelming, however, and can be simplified through preplanning, preparation, and practice. This chapter will provide some simple, straightforward guidance in the form of worksheets that may be adapted to your specific purpose. These worksheets may be used as is, or you may wish to add details specific to your application.

You will be guided through the following steps toward preparing a co-worker assisted rescue preplan, with suggested worksheets for each of the following:

Falls from Height: A Guide to Rescue Planning, First Edition. Loui McCurley.

1. Identifying responsible persons
2. Developing a policy statement
3. Performing a basic fall hazard survey
4. Mitigating fall hazards
5. Developing a fall protection plan
6. Developing a post-fall rescue plan
7. Coordinating with local agencies
8. Recordkeeping and training.

Let's get started!

STEP 1: IDENTIFYING RESPONSIBLE PERSONS

As part of your co-worker assisted rescue plan for workers at height, the employer should designate a program administrator for the fall protection plan. Rescue may be deferred to others as part of this plan, but the program administrator will be responsible for overseeing the plan in its entirety.

The first responsibility of the program administrator will be to record the names of those individuals responsible for preparing the plan and the names of those who are responsible for carrying it out. At the very least, the names of the program administrator, the person who prepared the fall protection plan, and the person who prepared the rescue plan should be recorded. If external consultants or other resources are used, this information should be recorded as well.

In addition to names of responsible persons, also include a statement of validity specifying the lifespan of the document. Any time the personnel or their responsibilities change, the plan should be updated accordingly. See Worksheet 11.1.

STEP 2: POLICY STATEMENT

The policy statement provides general goals and guidance for a managed fall protection program. It should emphasize management's commitment to providing a safe workplace for employees who may be exposed to fall hazards. A policy statement is not a generic statement, but should be customized to each employer. Your policy statement should reflect information as it is specific to your organization. The ANSI Z359.2 document, previously discussed as a guide to a comprehensive managed fall protection program, provides information on policy statements as foundational to developing your company's policy statement. Worksheet 11.2 will help you to prepare your policy statement for a post-fall rescue plan.

You may interview your employer, safety professionals, and workers to gather information by which to customize this statement for your organization. Include more detailed information on which employees will be protected, what constitutes an appropriate fall protection system, and in what situations fall hazards are known to exist in your organization.

Post-Fall Rescue Plan

Date ____________________

Employer Designated Program Administrator:

Name ____
Title ____
Company ______________________________________
Address ______________________________________
City ____________________ State ________________
Country ____________________ Zip ______________
Phone ___________________Alternate _____________
Email ____

Fall Protection Plan Prepared by:

Name ____
Title ____
Company ______________________________________
Address ______________________________________
City ____________________ State ________________
Country ____________________ Zip ______________
Phone ___________________Alternate _____________
Email ____

Rescue Plan Prepared by:

Name ____
Title ____
Company ______________________________________
Address ______________________________________
City ____________________ State ________________
Country ____________________ Zip ______________
Phone ___________________Alternate _____________
Email ____

Valid Dates:

This Plan is valid from _________ through __________ or until replaced by a more current plan, whichever occurs first. This plan will be updated whenever responsible persons, work practices, equipment, or other significant aspects of the program change.

Worksheet 11.1 Worksheet for recording responsible persons.

Worksheet for Preparing a
Post-Fall Rescue Policy Statement

Answer the following questions:

1. What is the name of the company?

   ____

2. Name the location/department (if applicable) to which this policy applies.

   ____

3. Which employees are exposed to fall hazards?

   ____

4. In what situations are these employees exposed to fall hazards?

   ____

5. What constitutes an appropriate fall protection system?

   ____

6. In the event of a fall, what methods will the employer use for rescue?

   ____

Now, summarize the answers to these questions into a policy statement, such as:

> Acme Manufacturing is committed to taking every reasonable precaution to protect the health and safety of employees. All employees are expected to maintain 100% fall protection when working above four feet. All employees who are working at height receive periodic training in personal escape and coworker assisted rescue techniques.
>
> The Northeast Widget division of Acme Manufacturing has painters working at height in Sector B, where personal fall arrest systems are used. In the event of a fall, the sector supervisor will call 911. Meanwhile the affected employee will assist himself to the extent he is able. A co-worker assisted response group consisting of at least four workers will be assigned for each shift, and will respond immediately to any incident in this sector.
>
> The Northeast Widget division of Acme Manufacturing also has engineering inspectors working at height in Sector M, where rope access is the primary means of access. These personnel will work in teams of two persons. In the event of a fall, the sector supervisor will notify 911. Meanwhile, the affected employee will attempt personal escape using standard rope access techniques. The affected employee's team mate will respond immediately to perform coworker assisted rescue using typical rope access partner rescue techniques.

Worksheet 11.2 Worksheet for developing a policy statement.

STEP 3: FALL HAZARD SURVEY

The fall hazard survey is normally conducted by an organization's safety professional and will identify all fall hazards associated with employers workplace/practices. Each hazard should be listed outright, and perhaps even assigned a numerical or coded "Hazard ID" for later reference. Hazards may be listed in detail as shown in Worksheet 11.3.

Fall Hazard Survey

Identify all fall hazards associated with employers workplace/practices
List each with a separate "Hazard ID"; add pages as necessary to accommodate multiple Hazard IDs

Hazard ID _____________ (Assign a reference number or name)

Is this a *Location Specific* or a *Multi-Location* Hazard?
(circle one)

Describe actual or typical location

Detail of hazard (diagram as necessary)

How is this location typically accessed? (consider preparing a separate Hazard ID for access)

Type of work occurring in this location

Persons potentially exposed (include number simultaneously exposed & duration of task)

Height of the potential fall (note obstructions)

Environmental Exacerbations: (check and describe all that apply)

- ☐ Sparks, flames, and heat-producing operations________________________
- ☐ Hazardous chemicals________________________
- ☐ Electrical hazards________________________
- ☐ Environmental contaminants________________________
- ☐ Sharp/abrasive surfaces________________________
- ☐ Moving equipment and materials________________________
- ☐ Unstable, uneven, and slippery walking/working surfaces________________________
- ☐ Unguarded openings________________________
- ☐ Climatic and weather factors;________________________
- ☐ Other notable factors________________________
- ☐ Foreseeable changes in any of these conditions________________________

Describe history of previous incidents in this (or similar) location:

Frequency of exposure: 1 2 3 4 5
Never.........................Continuous

Injury potential if a fall should occur: 1 2 3 4 5
None................................Fatal

Hazard priority:

___________ + ___________ = (The greater this value, the more critical is expedited post-fall rescue)
Exposure Frequency + Injury Potential

Worksheet 11.3 Fall hazard survey.

STEP 4: MITIGATING FALL HAZARDS

Once fall hazards are effectively identified, measures can be taken to mitigate each one. While it is not acceptable to take a generic approach to mitigating hazards, if there is one single method that would be appropriate for mitigating all hazards it is acceptable to adopt it. However, the validity and viability must be entertained for each known hazard individually.

Mitigation of fall hazards should be recorded in written form. While there are numerous ways to accomplish this, one example that clearly shows independent thought and consideration for each hazard is shown in Worksheet 11.4.

STEP 5: FALL PROTECTION PLAN

In this step, you will use the information gleaned in Worksheet 11.1, Worksheet 11.2, Worksheet 11.3, and Worksheet 11.4 as a foundation to create a fall protection plan where the method(s) of protecting the worker(s) is clearly outlined for each hazard identified. For each hazard identified, the fall protection system used in that location should be clearly specified, including diagrams. This will set the foundation for the rescue system to be utilized in the event of a fall at that location. An example of what such a plan might look like is outlined in Worksheet 11.5.

All equipment that is to be used in protecting against identified hazards should be specified in the plan. If particular performance criteria are required, this should be noted in the plan. This will lay the groundwork for recordkeeping, training, and inspection. Other components may also be recorded. These details should be updated if the make, manufacture, model, or other details change. Additional detailed information may be required for certain components. For example, in the case of rope(s)/ vertical lifelines, the employer may wish to specify a required terminated strength, noting what terminations are approved for use and who may be permitted to install terminations. Other criteria that the employer may wish to specify includes line material, diameter, and use of a stopper knot. Similarly, the employer may wish to specify detailed information regarding braking devices and rope grabs, such as fail-to-safe features, certification requirements, adjustability of friction, compatible rope diameters, and acceptable tie off method(s). As part of the fall protection plan, the program administrator should establish, utilize, and document some method of ensuring that all equipment purchased/replaced meets applicable regulatory requirements and standards.

STEP 6: POST-FALL RESCUE PLAN

Finally, with the fall protection plan in place, a plan for post-fall rescue can be established. Post-fall rescue practices should include guidance toward self-rescue (emergency escape), co-worker assisted rescue, and professional/municipal response.

Coordination with municipal resources is imperative, and contact information should be included in the response plan. Notes should also be included regarding

Fall Hazard Mitigation

Hazard ID ______________ (Correspond with hazards identified in Step 3)

1. **Elimination or Substitution.** Can the hazard or hazardous work practices be removed or can a substitute work method be used?
   < > Yes (On Fall Protection Plan page, describe how hazard is eliminated)
   < > No (explain why not and proceed to Question #2)

2. **Passive Fall Protection.** Can the the hazard or hazardous work practice be isolated or separated from employees or others? (Via guardrail, cover, etc.)
   < > Yes (On Fall Protection Plan page, describe how hazard is isolated)
   < > No (explain why not and proceed to Question #3)

3. **Positioning**. Can this fall hazard be mitigated through the use of a positioning system? (Maximum acceptable free fall is 2 feet.)
   < > Yes (On Fall Protection Plan page, describe positioning plan)
   < > No (explain why not and proceed to Question #4)

4. **Restraint.** Can this fall hazard be mitigated through the use of restraint systems? (Max slope 4:12)
   < > Yes (On Fall Protection Plan page, describe restraint system)
   < > No (explain why not and proceed to Question #5)

5. **Rope Access**. Can this fall hazard be mitigated through the use of a rope access system?
   < > Yes (On Fall Protection Plan page, describe rope access plan)
   < > No (explain why not and proceed to Question #6)

6. **Fall Arrest.** Can this fall hazard be mitigated through the use of a fall arrest system?
   < > Yes (On Fall Protection Plan page, describe fall arrest plan)
   < > No (explain why not)

NOTE: It is recommended that work that cannot be protected by one of the above methods be avoided altogether.

Worksheet 11.4 Fall hazard mitigation.

Fall Protection Plan

Hazard ID ______________ (Correspond with hazards identified in Step 3)

Overview
Based on the fall hazard survey for this particular identified hazard, and the fall hazard mitigation procedure selected, describe in detail and diagram the fall protection procedures to be used (use additional pages as needed):

How many workers will be at this specific site simultaneously: Minimum______Maximum_______

How many workers might simultaneously be exposed to this fall hazard? Minimum____Maximum_____

List individual members of the work team and their respective duties:

What safety provisions will be in place for nearby persons not on the work team?

Equipment
Describe all equipment to be used in protecting against this specific hazard.

Item	Make	Model	Certifications	Notes
Anchorage connector				
Harness (note approved Connection Points)				
Will a workseat be used?				
Lanyard(s)				
Rope grab				
Harness connector				
Spreader bar				
Rope(s)/vertical lifeline				
Rope adjuster(s) and/or descender(s)				
Other...				

Detailed information regarding rope(s)/vertical lifeline *(example)*

Approved terminations
Terminated strength
Diameter
Use of stopper knot
Other

Detailed information regarding rope adjuster(s) and/or descender(s) *(example)*

Fail-to-safe features
Certifications
Adjustable friction
Compatible rope diameter(s)
Tie-off method
Other

List supplemental work equipment (Equipment over12# should have a separate safety system)

Worksheet 11.5 Fall protection plan.

Installation

Describe and diagram anchorage locations (note permitted direction(s) of load)

Select One:

Certified Anchorage (min 2x MAF = ___________) x # Authorized Connections __________ = ____________
Restraint non-Certified (min 1000# (4.5kN)) x # Authorized Connections__________ = ____________
Positioning non-Certified (min 3000# (13.3kN)) x # Authorized Connections __________ = ____________
Rope Access non-Certified (min 5000# (22.2kN)) x # Authorized Connections __________ = ____________
Vertical Lifeline non-Certified (min 5000# (22.2kN)) x # Authorized Connections __________ = ____________
**note: Horizontal lifeline anchorages MUST be CERTIFIED by a QUALIFIED PERSON*

System Limitations

maximum free fall (2 ft / 6 ft)
fall clearance/obstructions
maximum arrest force (900lbf/1800lbf)
maximum number & permitted locations of authorized persons

Other notable protective measures to consider: (i.e., helmet, eyeprotection, footwear, environmental protection)

Installation/Dismantling

System to be installed no earlier than ___________ but no later than ___________

Party responsible for installation:

System to be dismantled no earlier than ___________ but no later than ______________

Party responsible for dismantle:

Provide detailed installation instructions as necessary:
(attach manufacturers instructions if applicable)

Describe any limitations on where or how the system shall be installed

Specific instructions for disassembling the fall protection system (if applicable)

Note where special instructions, MSDS, procedures for summoning rescue, and other notices must be posted on site:

Location where special instructions, notices, and written procedures for summoning rescue is to be posted:

Name of Qualified Person approving this plan:

Signed ____________________________________

Worksheet 11.5 (*Continued*)

any advance conversations with municipal resources, especially where expectations may have been established. Worksheet 11.6 provides a guide for preparing this document.

Summarized emergency contact information should be posted in a conspicuous location, and the employer should ensure that those who may be required to contact emergency services know how to do so.

Rescue Preplan
Overview

Local Municipal Resources

	Agency Name	Contact Person	Phone	How to Activate
Ambulance/Medical				
Nearest Hospital				
High Angle Rescue				

Identify what agency will be used for:

Self-Rescue	
Initial Response 5-45 minutes	
Professional Rescue 45+ minutes	
Expanded Capability	
First-Aid	

Complete the following for each hazard identified:

Hazard ID ______________
(refer to Hazard Survey)

What immediate action(s) should a worker take to notify co-workers in the event of a fall or other emergency?

What immediate action should co-workers take to notify company representatives, emergency services, or other authorities in the event of a fall?

What is the minimum and maximum amount of time it will take for responders from municipal agencies to reach the suspended worker?

	Minimum	Maximum
High Angle Rescue:		
Ambulance/Medical:		

Where can EMS agencies expect to access MSDs sheets and other specific information applicable to the worksite?

Complete the applicable section(s) below for the above incident

Describe in detail self-rescue procedures for a suspended worker (diagram as appropriate):
Equipment:

Methods:

These methods should be practiced (circle one) Monthly / Semi-Annually / Annually by all affected workers

Location(s) of on-site rescue cache:

Describe co-worker rescue procedures for a suspended worker from above using raising techniques (diagram as appropriate):
Equipment:

Methods:

These methods should be practiced (circle one) Monthly / Semi-Annually / Annually by all affected workers

Worksheet 11.6 Post-fall rescue plan.

Describe co-worker rescue procedures for a suspended worker from above using lowering techniques (diagram as appropriate):
Equipment:

Methods:

These methods should be practiced (circle one) Monthly / Semi-Annually / Annually by all affected workers

Describe co-worker rescue procedures for a suspended worker from above using descent techniques (diagram as appropriate):
Equipment:

Methods:

These methods should be practiced (circle one) Monthly / Semi-Annually / Annually by all affected workers

Describe co-worker rescue procedures for a suspended worker from above using lowering techniques (diagram as appropriate):
Equipment:

Methods:

These methods should be practiced (circle one) Monthly / Semi-Annually / Annually by all affected workers

Describe other approved, preplanned co-worker rescue procedures for a suspended (diagram as appropriate):
Equipment:

Methods:

These methods should be practiced (circle one) Monthly / Semi-Annually / Annually by all affected workers

Describe in detail professional rescue procedures for a suspended worker (diagram as appropriate):
Equipment:

Methods:

These methods should be practiced (circle one) Monthly / Semi-Annually / Annually by all affected workers

Worksheet 11.6 (*Continued*)

For each hazard identified in the original fall protection plan, define emergency response protocols in detail. These should include measures for emergency escape, co-worker assisted rescue, and professional response, as applicable. Repeat this exercise for each and every hazard identified.

This worksheet will serve as the foundation for training affected workers and co-worker assisted rescue responders, so be very specific and diagram as necessary.

Affected personnel should be trained in accordance with the written plan. If the plan, equipment, methods, techniques, or personnel change, retraining should take place accordingly.

STEP 7: COORDINATION WITH LOCAL AGENCIES

The capabilities of rescue services vary widely between jurisdictions, and not all agencies are capable of responding to all types of incidents. Municipal response organizations should at the very least be informed of the hazards that exist in your workplace, and, if possible, dialogue and/or even cooperative practice sessions should periodically take place to ensure adequate preparation on all fronts.

As a part of this process, a meeting with local area rescue/emergency services providers should be held to share information, set forth expectations, and to plan and coordinate the required evaluations. At this meeting, the employer should discuss key aspects of rescue with the organization, understanding that the goal is to gain understanding rather than to assume or insist that a certain capability be mandated.

In preparation for this discussion, the employer should understand his or her own needs with respect to the amount of time it may take for the rescue service to receive notification, arrive at the scene, and set up and be ready for rescue. The employer should be prepared to make clear to the response organization how to get from the entryway of the facility to the location or locations where rescue might most likely be necessary.

Likewise, the response organization should provide honest and straightforward information about their capabilities and limitations, including response time to the employer's workplace with respect to distance, roadways, and traffic. It is quite possible that the rescue service may predict different response times on different days of the week or at different hours of the day, or if it is already engaged in another response at the time of a call.

Finally, there should be open discussion of whether the rescue service does or does not own the equipment necessary to perform rescues, or if the equipment must be provided by the employer or another source.

Worksheet 11.7 may be used to help ensure that key points are addressed when consulting with potential rescue resources. Complete a separate sheet with each professional rescue agency potentially involved.

STEP 8: RECORD KEEPING AND TRAINING

The employer should designate a program administrator to be responsible for outlining and documenting all phases of the fall protection and post-fall rescue plan. At a minimum, program administrators should have a working knowledge of current fall protection regulations and standards as well as specific knowledge applicable to the industry and job in which they are performing.

Preplan Worksheet for Liason with Professional Rescue Agency

Use this worksheet to ensure that key points are addressed when consulting with potential rescue resources
Complete a separate sheet with each professional rescue agency potentially involved

Name of Agency	
Contact Person/Liason	
Office Phone	
Agency Activation Method	
Date of Meeting	
Worksite Address	
Covered Dates	

Meeting Attendees

Program Administrator:__________________________________

Agency Liason:______________________________________

Auditor:____

The following questions should be reviewed, mutually agreed, and signed off by all attendees.

	Administrator	Liason	Auditor
1. Agency is familiar with worksite and has reviewed fall hazard survey			
2. Agency is capable of responding to all potential falls covered by the fall hazard survey, with the following exceptions: (if none, write "none." Use addendum if needed.)			
3. Maximum response time to areas covered by the managed fall protection plan is _____ minutes.			
4. Agency is familiar with all fall protection equipment used on worksite.			
5. Agency must be notified _____hrs in advance when the following types of activities are planned:			
6. Agency will use the following methods to rescue suspended workers (use addendum if needed):			
7. Agency will maintain capability for this type of rescue by practicing on at least a (circle one) weekly / monthly / quarterly / annual basis.			
NOTES:			

Signed:

Program Administrator:____ Date:________

Agency Liason:____ Date:________

Auditor:____ Date:________

Worksheet 11.7 Coordination with local agencies.

Program Administrator Training

Program Administrator Prerequisites: **List Applicable OSHA 10/30 Hour courses:**

Other ____________________

At a minimum, program administrators should have a working knowledge of the following current fall protection regulations and standards as applicable to this job:

At a minimum, program administrators should be trained in the following fall protection concepts, equipment, and systems applicable to fall hazards experienced in this job (Based on fall protection plan; Include timeline recommendations):

At the end of the training, program administrator should possess the following knowledge and skills:

- Familiar with all skills expected of an Authorized Person working under him
- Use and administration of this managed fall protection program
- Use and maintenance of company's fall hazard survey
- Selection and appointment of safety committee personnel
- Fall protection system selection in accordance with this program
- Development of approved equipment purchase lists
- Selection and appointment of competent and qualified people
- Administration of rescue plans and procedures outlined by this document
- Administration of training programs outlined by this document
- Administration of accident / incident / near miss investigations

Training Records
List or attach records of all applicable training classes this person has taken. Training records should include, at a minimum:

Student's name; training organization name; trainer's name; evaluating organization's name; evaluator's name; dates and times of training; dates and times of evaluations; course objectives and content of education program; performance record of student based upon observation of physical demonstrations of skill or on theoretical exercises.

Recommended Frequency of Refresher Training:

Fall protection methods ______Months
Equipment inspection methods _______Months
Rescue methods _______Months
Administrative skills ________Months

Worksheet 11.8 Program administrator training.

Program Administrator Training

The employer should specify qualifications and training for the program administrator, including specific criteria for fall protection concepts, equipment, and systems applicable to fall hazards experienced in the job. These criteria should be based on the types of fall protection and other hazards that they are likely to experience in that position. A worksheet for developing program administrator training guidelines for your organization can be found in Worksheet 11.8.

Authorized Person Training

All employees who might possibly be exposed to fall hazards how to recognize and avoid them. Beyond this, those who might be authorized to enter a fall hazard zone

(Authorized Person) should receive more specific fall protection training. A prerequisite to entering any potential fall hazard should be completion of an approved OSHA 10-hour fall protection familiarization class. This class may be provided by internal or external resources. Specifying a minimum duration for training courses or curriculae cannot ensure adequate knowledge. There is no harm in including minimum timeline recommendations, but in addition, training content and evaluation criteria may be also specified.

Authorized Persons should be trained in fall protection concepts specific to fall hazards that may exist in the job that they might be expected to perform. Such training should be based on the defined fall protection/rescue plan for the area where they will be working and should at least address applicable aspects of the following concepts:

1. The specific type and characteristics of fall hazards in the work area;
2. Correct procedures for assembling, using, maintaining, and inspecting fall protection systems;
3. How to work in and around various hazards and protective measures, including controlled access zones, guardrails, personal fall arrest, safety nets, warning lines, and safety monitoring systems;
4. The employee's role within the employer's fall protection plan(s) and safety monitoring system(s);
5. Limitations of fall arrest equipment when used in certain applications, such as low-slope roofs.

Worksheet 11.9 will lead you through a process for developing an Authorized Person training program. This worksheet will help you create a big picture plan, but consider customizing the training program for Authorized Persons in different circumstances.

Rescue Training

In addition to fall protection-specific training, co-worker assisted rescue responders must also be trained to carry out the established emergency plan. The employer must periodically evaluate the capability of assigned employees to perform to the necessary levels in the event of a real emergency. In other words, the plan must be verified to realistically correspond to the potential need within a given workplace or environment.

A training program for co-worker assisted rescue responders should include guidance and hands-on practice for:

1. Sizing up existing and potential conditions at an incident, including scope and magnitude of the incident, information about the physical and mental condition of the subject, what resources will be required, how long the operation can be expected to take, and environmental factors.
2. Initial and ongoing hazard assessment.

Authorized Person Training

***** train <u>all</u> employees to recognize fall hazards *****
*****direct <u>all</u> employees not to expose themselves to fall hazards without fall protection!!*****

Authorized Person Prerequisites: List applicable OSHA 10/30 Hour Classes

Other ______________________

At a minimum, Authorized Persons should be trained in the following fall protection concepts specific to fall hazards experienced in this job: (Based on fall protection plan; Include timeline recommendations):

PPE use and inspection training should include at a minimum:

At the end of the training, Authorized Person should be capable of the following skills:

Training should include the following aids/methodologies
(List applicable manuals, equipment, audio / visual, physical, environmental training aids)

Recommended student-to-in structorratio

Recommended method(s) of evaluation

Minimum performance requirements of students

Training records should include:
Student's name; training organization name; trainer's name; evaluating organization's name; evaluator's name; dates and times of training; dates and times of evaluations; course objectives and content of education program; performance of student based upon observation of physical demonstrations of skill or on theoretical exercises.

Recommended Frequency of Refresher Training:
Fall protection methods ______Months
Equipment inspection methods _______Months
Rescue methods _______Months

Worksheet 11.9 Authorized Person training.

3. Methods for effecting the rescue without imposing undue hazards to rescuers or bystanders.
4. Activating the next level of emergency response.
5. Demobilization and debrief of involved resources.
6. Documentation.

Each member of a co-worker assisted rescue team should be provided opportunity to practice performing simulated rescues at least once every 12 months. In high risk environments, practice should be even more frequent than this. Practice should

Coworker Assisted Rescuer Training

***** train <u>all</u> Authorized Persons in personal escape *****
*****direct employees not to attempt to rescue another unless so authorized!!*****

Authorized Person Prerequisites: Fall Protection Training
Other ____________________

At a minimum, Authorized Persons should be trained in the following rescue concepts specific to potential rescue scenarios:

Knowledge of rescue equipment should include at a minimum:

At the end of the training, coworker assisted rescuer should be capable of the following skills:

Training should include the following aids/methodologies
(List applicable manuals, equipment, audio / visual, physical, environmental training aids)

Recommended student to instructor ratio

Recommended method(s) of evaluation

Minimum performance requirements of students

Training records should include:
Student's name; training organization name; trainer's name; evaluating organization's name; evaluator's name; dates and times of training; dates and times of evaluations; course objectives and content of education program; performance of student based upon observation of physical demonstrations of skill or on theoretical exercises.

Recommended frequency of refresher training:
Fall protection methods ______Months
Equipment inspection methods _______Months
Rescue methods _______Months

Worksheet 11.10 Co-worker assisted rescue training.

involve removing manikins or actual persons from environments and structures that closely resemble those from which rescuers might be required to perform an assisted rescue in an actual emergency. Guidance for developing a co-worker assisted rescue training plan is provided in Worksheet 11.10.

As part of each practice session, the practice rescue should be critiqued either by a knowledgeable internal resource or a qualified third party. The purpose of the critique is to identify deficiencies in procedures, equipment, training, or number of personnel so that they may be corrected. The results of the critique, and the corrections made to respond to the deficiencies identified, should be reviewed by the program administrator.

The following questions will assist employers in evaluating the performance of authorized co-worker assisted rescuers as well as external resources.

1. Does the employer's preplan adequately address all types of co-worker assisted response operations that may be likely to occur?
2. Do all responders demonstrate competency in performing their respective roles in accordance with the plan?
3. Do all responders demonstrate ability to recognize and care for signs, symptoms, and consequences of suspension intolerance and other hazards likely to exist in the workplace?
4. Is each potential responder provided with, and properly trained in the use of and need for, PPE and rescue equipment that may be needed during rescue operations?
5. Are responders adequately trained in basic first aid, cardiopulmonary resuscitation (CPR), and intervention for other specific medical conditions that may be specific to the facility?
6. Do rescuers perform their roles safely and efficiently with consideration first to their own safety and subsequently to that of the victim?

Equipment Inspection

Fall protection equipment should be inspected before each use. In addition, a more thorough, periodic inspection should be performed by a competent inspector. Such inspection should be recorded in a manner consistent with the employer's record-keeping methods. The employer should decide how often each component should be inspected, how the information should be recorded.

Inspection procedures and retirement processes should be outlined for each component. Worksheet 11.11 will guide you in developing appropriate inspection protocols for your program. Inspection and retirement should, at a minimum, be in compliance with manufacturer's instructions. The employer may set out additional criteria based on known workplace hazards or other conditions. Authorized Persons should be trained in accordance with the employer's designated practices.

Fall Protection Audit

The employer's fall protection program, including the post-fall rescue plan, should be periodically audited to ensure that it is up to date and that it is being effectively implemented. An audit may be performed by the program administrator or, for a more objective perspective, an external resource may perform the audit.

An audit should include full review of all applicable procedures, complemented by some means of verifying that employees are actually adhering to the process during the regular working days. Recommended audit intervals are at least once every 2 years, whenever Authorized Persons or Competent Persons identify deficiencies, or when there are incidents involving injuries, property damage, or "near misses." Worksheet 11.12 offers a format for generating a report from the audit.

Fall Protection Equipment
Inspection Protocols

Fall Protection Equipment Inspection

Recommended Inspection Intervals
(for each item check all that apply)

	Before each use	6 month periodic	Annual Periodic	Other
Anchorage				
Lifeline				
Lanyard				
Force Absorber				
Rope Grab				
Descender				
Harness				
Connectors				
Other: (list)				

Inspection Procedures

	Guidance (Consider visual and tactile methods. Include manufacturers recommendations and local protocols)
Anchorage	
Lifeline	
Lanyard	
Force Absorber	
Rope Grab	
Descender	
Harness	
Connectors	
Other: (list)	

Retirement Procedures

Worksheet 11.11 Equipment inspection.

Fall Protection Audit

Employer Name	
Employer Designated Program Administrator	
Plan Implementation Location	
Auditor (Company & Contact Information)	
Auditor Representative	
Date(s) Audit Performed	

Auditor should answer the following questions after reviewing the organizations Managed Fall Protection Program, interviewing selected competent persons and authorized persons, and observing the worksite.

Yes	No	
		Does employer have a documented managed fall protection program?
		Does employer's managed fall protection program include a written policy regarding fall protection?
		Is the policy applied in the workplace? Explain:
		Does employer's managed fall protection program identify the program administrator?
		Does employer's managed fall protection program include an adequate fall hazard survey to identify all reasonably foreseeable fall hazards?
		Does employer's managed fall protection program include a fall hazard mitigation plan in which fall hazards are considered in light of an appropriate hierarchy of controls?
		Does employer's managed fall protection program include a specific fall protection plan that addresses equipment, systems, and methods for protecting falls?
		Does employer's managed fall protection program address training considerations for all authorized persons?
		Is the training program for the required personnel adequate to perform their duties and responsibilities in light of the fall hazard survey? If not, explain:
		Does employer have training records on file to substantiate that training has been provided and individuals have been evaluated in accordance with the established program?
		Is there a plan to ensure that authorized persons are adequately trained and appropriately evaluated prior to starting work?
		Is the established training program in conformance with ANSI Z490.1?
		Does personal protective equipment used in the program meet applicable standards? Which standard(s)?
		Does the purchasing program ensure that future equipment purchased will also conform with applicable standards?
		Are fall protection systems in use adequate for their intended purpose?
		Are identified anchorage systems adequate?
		Is there an adequate procedure in place for inspection, maintenance, storage and care of fall protection equipment?
		Are the fall protection systems identified in the Plan being used as appropriate by exposed Personnel?
		Does employer's managed fall protection program include rescue procedures appropriate to the potential need?
		Are exposed employees adequately trained in established Rescue procedures?
		Does employer have records of meetings with local emergency services/rescue resources to identify capabilities and working relationships?

Worksheet 11.12 Fall protection audit form.

		Has employer experienced any incidents requiring any rescue (self-rescue, coworker rescue, professional rescue) in the last 5 years?
		Does employer have a record of incident reports from any falls that might have occurred in the last 5 years? Review and comment on resolution:
		Has the employer made insurance claims resulting from a fall in the last 5 years? Review and comment on resolution:
		Are particular strengths noted in the program? Explain.
		Are particular weaknesses noted in the program? Explain.
		List procedures, workplace activities, and/or equipment that should be considered for improvement:

ADDITIONAL NOTES:

It is the responsibility of the program manager to develop a documented plan of action to address issues found during this audit. The plan should identify what changes will be made, who is responsible for making each change, and when the change will be completed.

Consideration should be given to a team approach when developing recommendations, specifically including management and employee involvement.

This audit was performed on the ____ day of ________, 20__ by ____________________________

who is employed by ____________________________.

Signed____________________________________ (Auditor)

This audit, and resulting recommendations, was reviewed and accepted by employer's representative (program manager) ____________________________ on the ____ day of ________, 20___.

Signed____________________________________ (Program Manager)

Retain this audit documentation for at least 5 years or as required by law.

Worksheet 11.12 (*Continued*)

INCIDENT INVESTIGATION

Any incident or near miss should be thoroughly evaluated and considered with an eye toward preventing accidents in the future. Investigations should include at least the information shown in Worksheet 11.13.

Incident Investigation/Report

Report Prepared by: Date:

Date of incident:	Time of incident:
Location where incident occurred:	
Program administrator:	
Competent Person (supervisor) on duty at time of incident:	
List all individuals on site at the time of the incident: Name	Contact Phone #

Please review and respond to all of the questions in the following list, expounding as appropriate.

Yes	No	
		Was work shut down at the immediate location following the incident? How and by whom?
		Has all equipment involved in the incident been quarantined? Where:
		Did employer have an active, documented managed fall protection plan/policies in place at the time of the incident? (attach a copy to this report)
		Was the plan appropriate to the need? Explain:
		Had employee(s) been trained in accordance with the plan? Note date(s) of training, or if not, why not?
		Was employee(s) working in accordance with the plan at the time of the incident? If not, why not?
		Was there equipment failure involved in the incident? Explain:
		Was equipment failure caused by misuse (i.e., using product outside manufacturers recommendations)? Explain:
		Were any unsafe operations observed, reported, or present at the time of the incident? Explain:
		Were any unsafe operations observed or reported to be present at the time of the incident? Explain:
		Were any tools in use at the time of the incident? Explain:
		What equipment (PPE) was in use at the time of the incident? Explain:
		Was there any other equipment in use at the time of the incident? Explain:
		Are any defects in the design of the fall protection system or other surroundings noted? Explain.
		Was construction of the fall protection system and other involved structures appropriate? Explain.
		Was planning for this job and the related fall protection needs adequate? Explain.
		Was the layout of the fall protection system, including access and egress, appropriate for the situation? Explain.

Worksheet 11.13 Fall incident report.

		Was appropriate rescue provision made and executed in this situation? Explain.
		Were there outstanding environmental/physical factors that may have contributed to this incident (i.e., noise, vibration, temperature extremes, radiation, and illumination) ? Explain.
		Were there outstanding ergonomic factors that may have influenced or contributed to this incident? Explain.
		Were there outstanding chemical or biological factors that may have influenced or contributed to this incident? Explain.
		Was there a lack of knowledge or skill that may have influenced or contributed to this incident? Explain.
		Was there a lack of management involvement that may have influenced or contributed to this incident? Explain.
		Could a lack of communication have influenced or contributed to this incident? Explain.
		Were any inadequate procedures in place at the time that may have influenced or contributed to this incident? Explain.
		Was there a lack of enforcement of procedures that may have influenced or contributed to this incident? Explain.

Other Notes:

Incident report created by ____

Signed________________________________ on the _____ day of _________, 20

Report received by program manager____

Signed ________________________________ on the _____ day of _________, 20___

Retain this incident report for at least 5 years or as required by law.

Worksheet 11.13 (*Continued*)

SUMMARY

By understanding the information in this book, and following the processes outlined in this chapter, you will be well on your way toward appropriate preparation for rescue. Completing the processes outlined in this book is just the beginning. Ongoing re-training, practice, inspection, and recordkeeping are integral to a successful program. Establish internal processes for ensuring these to help your program retain its integrity.

INDEX

Falls from Height: A Guide to Rescue Planning, First Edition. Loui McCurley.